Raimund Hoffer, William Theodore Brannt

A practical Treatise on Caoutchouc and Gutta Percha

Comprising the Properties of the raw Materials and the Manner of mixing and

working them

Raimund Hoffer, William Theodore Brannt

A practical Treatise on Caoutchouc and Gutta Percha
Comprising the Properties of the raw Materials and the Manner of mixing and working them

ISBN/EAN: 9783337176877

Printed in Europe, USA, Canada, Australia, Japan

Cover: Foto ©ninafisch / pixelio.de

More available books at **www.hansebooks.com**

A

PRACTICAL TREATISE

ON

CAOUTCHOUC AND GUTTA PERCHA;

COMPRISING

THE PROPERTIES OF THE RAW MATERIALS AND THE
MANNER OF MIXING AND WORKING THEM;

WITH

THE FABRICATION OF VULCANIZED AND HARD RUBBERS,
CAOUTCHOUC AND GUTTA PERCHA COMPOSITIONS,
WATERPROOF SUBSTANCES, ELASTIC TISSUES,
THE UTILIZATION OF WASTE, ETC.

FROM THE GERMAN OF
RAIMUND HOFFER.

BY
WILLIAM T. BRANNT,
GRADUATE OF THE ROYAL AGRICULTURAL COLLEGE OF ELDINA, PRUSSIA.

ILLUSTRATED.

PHILADELPHIA:
HENRY CAREY BAIRD & CO.,
INDUSTRIAL PUBLISHERS, BOOKSELLERS, AND IMPORTERS,
810 WALNUT STREET.

LONDON:
SAMPSON LOW, MARSTON, SEARLE & RIVINGTON,
CROWN BUILDINGS, 188 FLEET STREET.
1883.

TO

N. CHAPMAN MITCHELL, Esq.,

OF PHILADELPHIA,

This Volume

IS RESPECTFULLY DEDICATED,

BY HIS FRIENDS,

THE PUBLISHERS.

PUBLISHERS' PREFACE.

THE great and growing importance and magnitude of the Industries of Caoutchouc and Gutta Percha, especially in this country, and the almost entire absence of any literature on the subject in the English language, together with the urgent desire of their friend N. CHAPMAN MITCHELL, Esq., of this city, that such a publication should be made, have been the causes which have induced the publishers to issue this volume.

The German work upon which the present one is based is an admirable one, and has been carefully and conscientiously translated and edited by Mr. Brannt, and it is believed that the result is a treatise which will be found to be a valuable and creditable addition to American Technical Literature, as well as a useful handbook for those who are engaged in the importation, sale, and purchase of the raw materials, and the manufacture of all classes of goods from these materials, and the utilization of the waste products. With this feeling it is confidently submitted to the Public.

PHILADELPHIA, Dec. 1, 1882.

CONTENTS.

I. Introduction.
	PAGE
Caoutchouc; Gutta Percha	25
Earliest information about Caoutchouc; first correct data regarding Caoutchouc due to French explorers	26
De la Commandine, Fresneau, Priestly; Early use of Caoutchouc	27
Cultivation of Caoutchouc throughout the tropical zone; Principal localities for producing Caoutchouc; Mazer-Wood; Gutta Percha, its first introduction into England by Dr. Wm. Montgomery	28
Observations of Dr. Montgomery in the East Indies regarding Gutta Percha; Study of the properties of Gutta Percha	29
Behavior of Caoutchouc and Gutta Percha nearly the same; Balata and Coorongit; Jintawan	30

II. Where Caoutchouc is Found, and how it is Gained.

Caoutchouc found in the milky juices of plants; Found in some European plants; Families of plants containing trees whose milky juice contains Caoutchouc in large quantities	31
Trees which furnish the larger quantity of Caoutchouc used; Ficus elastica	32
Urceola Elastica, the most prominent of the Caoutchouc-furnishing trees in the East Indies; Siphonia Elastica, the American Caoutchouc tree; Enormous extent of the import of Caoutchouc; How Caoutchouc is obtained	33

III. Commercial Caoutchouc.

	PAGE
Classification and Value of Caoutchouc; A.—American Caoutchouc; Carthagena Caoutchouc.	37
West Indian Caoutchouc; Para Caoutchouc.	38
B.—East Indian Caoutchouc; Bombay, Calcutta, etc. C.—African and Madagascar Caoutchouc; Angola, Benguela, Congo, and Gaboon	39

IV. Properties of Caoutchouc.

A.—Physical Properties.	40
Specific Gravities of different samples of Caoutchouc as well as of the same samples under different conditions.	43
B.—Chemical Properties	44
C.—Behavior towards Sulphur	46
D.—Behavior towards Solvents	47
Average results of Experiments in dissolving Caoutchouc with different Solvents	49
Of what Caoutchouc consists; on what Elasticity of Caoutchouc depends; on the complete dissolution of Caoutchouc	50
The use of Oil of Turpentine for dissolving Caoutchouc; dissolving Caoutchouc in boiling Linseed Oil	51
The so-called Oil of Caoutchouc; The most suitable Solvents for Caoutchouc.	52
Fry's Patent Process for dissolving Caoutchouc	53
E.—Behavior in Heat; bodies which have been found as existing in the Oil of Caoutchouc	54
The Preparation of Oil of Caoutchouc as a Solvent for Caoutchouc.	56

V. Mode of Working Crude Caoutchouc.

	PAGE
Freeing Caoutchouc from Foreign Admixtures	56
Importance of using the Crude Material of one variety at a time	57
Machine for Cutting up Caoutchouc	59
To obtain Thin and Tender Shavings	60
The Hollander and its Use	61
Kneading Rollers	63
Conditions under which the Homogeneity of the mass treated in the Kneading Machine increases; Moulds for shaping Caoutchouc that comes from the Kneading Machines	65
Machines for the treatment of Caoutchouc, which are to be preferred; To obtain the Caoutchouc in a homogeneous mass; Advantages of storing Caoutchouc after the preparatory preparation; Methods of further treating the material	66
Vulcanizing or Sulphurizing	67

VI. Preparation of Vulcanized Caoutchouc or Vulcanite.

Effect of Temperature on the Elasticity of Caoutchouc	67
Great importance of the discovery of the Vulcanizing Process; Properties given by Vulcanizing	68
Hard Rubber; the discoveries of Lüdersdorf and Goodyear	69
Various methods by which Sulphur may be introduced; Substances which are mixed with Vulcanized Caoutchouc	70
Properties of Vulcanized Caoutchouc	71
Experiments of Payen and others; effects of an excess of Sulphur on the Elasticity of Caoutchouc	72
Desulphurized Rubber	73

VII. Vulcanizing with Pure Sulphur.

	PAGE
Temperature required	73
To obtain an entirely Uniform Product; Burning	74
Temperature at which Chemical Combination takes place	75
A.—Mechanical Combination of Caoutchouc with Sulphur	76
The use of finely powdered Rose Sulphur or Flowers of Sulphur; Mode of Proceeding	77
B.—Burning of the Caoutchouc and Sulphur mass	78
The Time required and the degrees of Heat	79
The Precautions necessary in the construction of the Apparatus	80
C.—Burning Apparatuses	83
D.—The Burning Operation	84

VIII. Vulcanizing with the Assistance of Chloride of Sulphur.

Parkes' Cold Vulcanizing Process; Gaulthier de Caulbry Process	87
Precautionary measures for the Protection of the Workmen	89
Preparation of Chloride of Sulphur	91
Refining of Petroleum	93

IX. Use of Sulpho-Metals for Vulcanizing.

The Sulphides of the Alkali Metals well adapted for the purpose	94
Vulcanized masses; Coloring	97
Deodorizing Vulcanite	101
Desulphurized Vulcanite	102

X. Preparation of Hard Rubber.

Cornite and Keratite; Uses of Hard Rubber	104
Large quantity of Sulphur used in making Hard Rubber; Zinc Penta-Sulphide; Mixing the Ingredients	105
Burning; Temperatures used in Burning	106

	PAGE
Coloring	108
Dusting; Plating; Enamelling	109
Coloring substances which cannot be used in the manufacture of Hard Rubber articles; Use of Lake and Zinc Colors	110
Proportions of constituents to be used in the manufacture of Hard Rubber	111
Constituents which must not be used in making Hard Rubber which is to resist Chemical Influences; Use of Hard Rubber	113

XI. Preparation of Artificial Ivory, Ebonite, Eburite, or "Ivoire Artificiel."

Impossibility of Bleaching Caoutchouc without producing chemical change	115
Apparatus for treating Caoutchouc with Chlorine	116
American Bleaching Process	117
Simplest Method of Bleaching Caoutchouc for preparing Ebonite	118
Other Processes; Steps necessary in the Manipulations	119
An American receipt with Constituents of an Ebonite mass	120

XII. Caoutchouc Compositions.

Kamptulicon	121
Caoutchouc Leather	123
Balenite	124
Plastite	125
Grinding and Polishing Compositions, with Receipts	127
Caoutchouc Enamel	130

XIII. Caoutchouc Lacquers.

Preparation of Caoutchouc Solutions	133
Apparatus adapted for Kneading Caoutchouc Solutions	135

	PAGE
Caoutchouc Varnish for Leather	137
Caoutchouc Varnish for Gilders; Caoutchouc Varnish for Glass; Marine Glue; Marine Glue for Damp Walls	138
Jeffery's Marine Glue; Hard Rubber Lacquer	140
Caoutchouc Cements; Cements for Glass	141
Cement for Rubber Shoes	142
Soft Cement of Caoutchouc and Lime	143
Hard Caoutchouc Cement; Gutta Percha Cements; for Glass and for Leather	144
Cement for Rubber Combs; Elastic Gutta Percha Cement	145
Asphaltum Cement for Leather Straps; Gutta Percha Cement for Horses' Hoofs	146
Heveenoid; Soft Heveenoid	147
Hard Heveenoid; Metallized Caoutchouc; Cheap Erasing Rubber	148

GUTTA PERCHA.

Gutta Percha, whence derived, and characteristics of . 150

XIV. PROPERTIES OF GUTTA PERCHA.

A.—Physical Properties	152
Behavior of Gutta Percha at different temperatures	153
B.—Chemical Properties	155
Payen's Analysis of Gutta Percha	156
Albane and Fluavile, and Gutta and their relations to each other; other combinations in crude Gutta Percha	157
Clark and Miller's Experiments upon Gutta Percha	158
Miller's examination of Cables which had been submerged	159

XV. CLEANSING OF CRUDE GUTTA PERCHA.

Machines for Cutting Gutta Percha	162
Cleansing of Shavings	163

XVI. Vulcanization of Gutta Percha.

	PAGE
Qualities of Vulcanized Gutta Percha	165
Practical Operation of Vulcanizing; Temperature in Burning	166
Simplest Plan of Vulcanizing Gutta Percha; Bisulphide of Carbon for coating articles with Vulcanized Gutta Percha	167

XVII. Bleaching of Gutta Percha.

The Bleaching Processes	168
Apparatus	169

XVIII. Gutta Percha Compositions.

Gutta Percha and Caoutchouc Composition for Machine Belts	172
Hard Gutta Percha Compositions	174
Compositions of Gutta Percha and Wood	176
Sorel's Gutta Percha Composition	177
Rousseau's Method of Preparing Gutta Percha Solutions and their Use	179

XIX. Manner of Working Caoutchouc and Gutta Percha.

Exhibit of the American Division of the Paris Exhibition	181

XX. Manufacture of Caoutchouc and Gutta Percha Plates.

Cutting Plates from Blocks	184
Cutting Plates from Cylinders	185
Manufacture of Plates from Solutions; Sollier's process	187

	PAGE
Preparation of Plates by Rolling; Calenders	189
Stretching Machines	191
Preparation of Plates from Vulcanite Masses	193

XXI. Mode of Preparing Caoutchouc and Gutta Percha Threads.

Square Cords from Crude Caoutchouc	196
Manner of Cutting Square Cords from Vulcanized Caoutchouc	199
Round Caoutchouc Threads	200
Preparation of Gutta Percha Threads	203

XXII. Fabrication of Caoutchouc and Gutta Percha Hose or Tubing.

A.—Caoutchouc Hose	205
Fabrication of Ordinary Hose	206
Hose with Inclosures	208
B.—Gutta Percha Hose	209

XXIII. Forming or Moulding of Massive Articles and Hollow Bodies from Caoutchouc and Gutta Percha.

Complicated Articles; Dolls and Toys	214
Elastic Balls	216
Balloons	218
Coating of Wires with Gutta Percha	219
Apparatus	221
Length, etc., of some of the Submarine Cables in use	223

XXIV. Fabrication of Caoutchouc Sponges.

Practical Operation	224
Disagreeable Odor of these sponges, and the mode of removing it	225

XXV. Fabrication of Rubber Shoes.

XXVI. Fabrication of Waterproof Tissues with Caoutchouc.

	PAGE
Mackintosh's Invention	227
Dumas's Invention	228
Tissues to be used for Waterproofing	229
The more Modern Processes	230
Spreading the Caoutchouc Mass	233
Apparatus	234
Apparatus for Protecting the Workmen from the effects of the vapors and for Spreading the mass Uniformly on the tissue	236
Manufacture of Tissues with Caoutchouc Inclosures; Apparatus for Uniting the Tissues	239
Deodorization of Water-proof Fabrics	240

XXVII. Manufacture of Water-proof Tissues by Means of Caoutchouc Compositions.

Substitutes for part of the Caoutchouc 242

XXVIII. Fabrication of Elastic Webbings.

Manufacture of the Caoutchouc Stamps . . . 247

XXIX. Waste and its Utilization.

XXX. Adulterations of Caoutchouc and Gutta Percha.

Nature of the Impurities; *Getah Malabeœga* . . 253
Tests for Getah 254

XXXI. Examination and Imitation of Caoutchouc and Gutta Percha Compositions.

	PAGE
Chemical Analysis	257

APPENDIX.

Statistical Data Respecting the Consumption of Caoutchouc and Gutta Percha and the Production of these Industries.

Imports into England and France	260
Imports into Hamburg	261
Statistics of the United States	262
Substitutes for Caoutchouc and Gutta Percha	265
Substitute for Gutta Percha	268
Cement for Vulcanized Caoutchouc; New Rubber Compound	269
Index	271

PRACTICAL TREATISE

ON

CAOUTCHOUC AND GUTTA PERCHA.

I.

INTRODUCTION.

AMONG the numerous products of the animal and vegetable kingdoms sent to us from the tropical regions since opening of the sea-route to India, and discovery of America, the substance known as caoutchouc, gum elastic, or India-rubber, occupies a prominent position as an indispensable auxiliary to many manufactures, and for its own development into an extensive industry.

The substance designated in commerce as gutta percha has been practically known to us only since 1843. Its introduction incited manufacturers with great hopes of its useful application to industrial pursuits. Although these anticipations have been but partially realized, nevertheless must gutta percha, for certain purposes, be ranked as an agent of great value to industry, and as one difficult to be replaced. Caoutchouc and gutta percha (belonging to the vegetable kingdom) are secreted from the milky juices furnished in remarkably great abundance by many plants. Although many milky juices contain caoutchouc, it would be erroneous to suppose they all do. In the milky juices from which caoutchouc is obtained are found small particles of caoutchouc, resin, fat, coloring matter, etc., which give the juices their peculiar milky appearance.

In addition to the ingredients mentioned, which are utilized in various ways, milky juices of certain plants (*Euphorbiaceae*) contain virulent poisons, used by the negroes to poison their war implements.

Our first information about caoutchouc, and the plants from which it is obtained, came to us from South America, where it seems that natives of certain parts had been for a long time familiar with the mode of obtaining and applying it. Travellers found, in some parts of Brazil and Guiana, utensils and shoes of caoutchouc in general use by the natives. Caoutchouc was first brought to Europe as a curiosity by travellers who had explored the tropical regions of South America. They brought it in the condition as prepared by the Indians, in round or oval-shaped bottles, but nothing definite was known of this remarkable substance. Some declared it to be of a vegetable origin, while others asserted it as belonging to the animal kingdom.

The merit of first giving correct data of caoutchouc

belongs to French explorers. *De la Commandine*, in 1735, positively declared it to be the desiccated juice of a tree indigenous to Brazil.

In 1751, *Fresneau* discovered a caoutchouc furnishing tree in Cayenne, and we are indebted to this indefatigable explorer for the first exact description of the method employed by the natives in obtaining it.

The celebrated English chemist *Priestly* was the first to call attention to a useful application of caoutchouc by recommending it for effacing lead-pencil marks (1770).

Although the origin and several important properties of caoutchouc were known in the first quarter of the present century, it was rarely applied to industrial purposes, its principal use being to efface lead-pencil marks, and was brought into the market in the form of bottles under the name of India-rubber, the trade being almost exclusively in the hands of the English. After the chemical properties of caoutchouc became better understood, it was found that we had a substance, which, on account of its extraordinary elasticity and indifference to chemical influences, was well adapted for manifold purposes, and, through the discovery of vulcanized and hard rubber, we have learned to prepare new combinations, utilized in numerous ways, and for special uses could not be well supplied by any substitutes. The result of a closer examination into the merits of caoutchouc led to its extended employment as an industrial factor, and created such an increased demand that it became necessary to seek new sources of supply.

We have mentioned that the first information of caoutchouc came to us from South America. This was supplemented by the knowledge of an East Indian plant also furnishing it. Since then botanists have discovered a large series of plants furnishing the desired object.

Cultivation of caoutchouc plants extends over the entire tropical zone, and experiments in growing such trees have been crowned with complete success. While these trees flourish everywhere within the tropical zone, the tropical regions of America, East Indies, Java, Nubia, and Central Africa may be especially named as the principal localities producing them. The greater portion of the entire world's production of caoutchouc reaches England, where are to be found the largest manufactories for production of caoutchouc goods in Europe. France and Germany come next to England. And the consumption of caoutchouc is steadily increasing, as it is coming more and more into general use.

A substance called *mazer-wood* was imported into England from the tropical regions of Asia for a considerable time without coming into practical use. There is scarcely any doubt at the present time that this mazer-wood and gutta-percha are identical substances.

Dr. William Montgomery, who had resided for many years in Singapore, imported, in 1843, the first gutta percha into England, and endeavored to introduce it into general use.[1]

Chemical science had made such progress at that period that not much time was required for studying more accurately the physical and chemical qualities of gutta percha, and as soon as it became known that the properties of gutta percha resembled in a considerable degree those of caoutchouc, and in some particulars even surpassed the latter substance, the importation of gutta percha was entered into on a large scale, and it is again in England where the gutta percha industry was first established, and where it has since then been widely extended.

---

[1] During one of his travels in the East Indies, *Dr. Montgomery* entered into conversation with a Malay laborer. While talking, he observed the handle of a hoe, and he heard with surprise that its substance, however hard it appeared to be, could be softened by immersion into hot water, and could thereupon assume and preserve any desired shape. The experiment being immediately made, the assertion of the Malay was fully confirmed. On further inquiry, that excellent quality of the substance in question was found to have been long known among the natives of Java, where it was used for manufacturing canes and handles of whips, as well as of various other implements, and especially of knives and daggers. *Montgomery* sent various specimens to London, and called attention to the manifold uses of which the substance was capable. His words were more duly noticed than those of *D'Almerida*, who about ten years ago had sent a similar freight to the Asiatic Society in London. Gutta percha was at the same time also discovered by *Thomas Lobb*, who, in 1842 to 1843, made a botanical journey through the East India islands. —*Translator.*

It will be shown later on in our work that the behavior of caoutchouc and gutta percha is nearly the same, and we, therefore, commence our treatise with the longer known and more generally used caoutchouc, so as to avoid an unnecessary repetition of particulars which have been already explained.

In modern times two other bodies resembling caoutchouc and gutta percha have attracted the attention of chemists and technologists. These are the two substances to which the names *balata* and *coorongit* have been given. Balata became known in Europe about 25 years ago, and is used in England at present, even if only to a limited extent. With respect to its properties balata very closely resembles gutta percha, and it is perhaps very likely that both products are identical substances, although derived from different plants. At the present time balata is brought into commerce only from Guiana.

Coorongit was brought to Europe from South Australia, and was at first considered a substance nearly related to caoutchouc, but closer examination has shown it to be of mineral origin, and that it possesses certain resemblances to ozocerite or asphaltum.

Balata is of very little importance for industrial purposes, caoutchouc and gutta percha being the substances belonging to this group of vegetable products which alone at the present day attract the attention of technologists.

For the sake of completeness we will here mention a product which, under the name of *jintawan*, has been brought to England from the East Indies, but it is very likely that it is not a new substance, but identical with gutta percha.

II.

WHERE CAOUTCHOUC IS FOUND, AND HOW IT IS GAINED.

Caoutchouc, or India rubber, is found, as before mentioned, in the milky juices of many plants, and according to some botanists is a never wanting constituent of every milky juice. It is also found in some European plants, for instance, in some species of *Euphorbia* and in fig-trees, but in such small quantities that it would not be worth while to make an attempt of gaining it from them on a large scale.[1]

We here limit ourselves to a brief enumeration of the families of plants containing trees whose milky juice yields a sufficient quantity of caoutchouc to make it possible to gain it in large quantities. No doubt this enumeration will be incomplete, in so far as the more the tropical regions are explored, the more caoutchouc yielding plants will certainly be found.

---

[1] It is claimed that the milky juice of milkweed (*Asclepias*) yields four per cent. of caoutchouc. Some time ago a company was formed in Canada, and an attempt made to gain caoutchouc from milkweed, but we have not been able to learn with what success. — *Translator.*

The following families of plants are the principal ones which furnish milky juices rich in caoutchouc, and from which it is gained in great abundance:—

Artocarpaceae (Bread fruit trees),
Apocynceae,
Euphorbiaceae.

All plants from which caoutchouc is obtained may be classed in these families. Among the *Artocarpaceae* are several species of fig-trees which serve for the purpose. *Ficus elastica*, originally indigenous to the East Indies, was first discovered there in 1810, and is at the present time cultivated as a caoutchouc-tree in all tropical regions of Asia and in Nubia. This tree with the *Bahea gumnifera* and *Urceola elastica*, the latter two belonging to the *Apocynceae*, may be called the original caoutchouc producers of Asia. *Bahea gumnifera* was originally indigenous to the island of Madagascar, *Ureola elastica* to the South Asiatic islands of Borneo and Sumatra. *Siphonia elastica*, belonging to the *Euphorbiaceae*, is found especially in America; *Hancornia speciosa* in the valley of the Amazon in Brazil, as also *Cecropia peltata* and *Castilion elastica* in Mexico, the latter belonging to the *Artocarpaceae*.

The trees above enumerated furnish by far the largest quantity of caoutchouc used, but there are still several others which serve for the same purpose, but occur comparatively in such small quantities, that they are of little importance, at least for industrial purposes.

Upon the mainland of the East Indies the *Ficus*

elastica forms gigantic forests. The trunk of the tree reaches a height of more than 30 metres (98.4 feet), and, it is claimed, as much as 8 metres (?) (26.2 feet) in diameter, and the tree has further the peculiarity of sending air roots from the end of its branches into the ground, and thus in the course of time, the forest becomes an almost impenetrable thicket.

Urceola elastica is the most prominent of the caoutchouc furnishing trees in the East India islands; it grows very quickly, reaches a gigantic size and can furnish every year as much as 30 kilogrammes (66 lbs.) of caoutchouc. The true American caoutchouc-tree, *i. e.*, *Siphonia elastica*, grows also very rapidly and obtains a considerable size. When we consider the quantities of caoutchouc which are imported every year into Europe alone, we may form some idea of the immense number of trees used in obtaining caoutchouc. But, as may be seen from the statistics of importation of the last ten years, the quantities of caoutchouc likely to be used in the future will increase to an enormous extent, and the demand for this material can only be satisfied by especially planting and cultivating trees to furnish it, as is being done at the present time in America, Asia, and India.

Caoutchouc is only found in the milky juice contained in particular cells of the respective plants, as may be easily shown by a microscopic examination. When the plant is punctured, the milky juice exudes at once, and on exposure to the air gradually hardens into a cream-like mass. This milky juice, being the product of living organisms and composed of several bodies, has a different composition during the various stages of vegetation. Besides water and caoutchouc, are found in it vegetable albumen, resin, salts, and other combinations, which have not been thoroughly examined. There seems to be a considerable variation in the percentage of caoutchouc, depending on the locality where the tree grows, the age of the tree, and the period of vegetation. Milky juices have been found containing nearly 37 per cent. of caoutchouc, while others showed scarcely 20 per cent.

An incision is made penetrating the sap conductors of the tree, causing the juice to exude, which is collected in cavities cut in the lower part of the tree, or, still better, a pan is fastened on the trunk to catch the flow.

If the milky juice, which is white when it exudes from the tree, is left to itself, it soon assumes an appearance similar to that of cows' milk; generally the small drops of caoutchouc rise to the top and form a cream-like layer floating upon a residue, which contains no more caoutchouc, and which also becomes decomposed when exposed to a higher temperature.

The labor of gathering caoutchouc, as long as it was a limited business, was done by the Indians by means of balls of clay,—about the size of a child's head,—fastened to wooden strips and coated with the creamy mass which had been separated from the milky juice. The coating was then dried over a densely smoking fire (by which it assumed a brownish to blackish color), and this process of coating and drying was repeated until the coating was from 4 to 5 centimetres (1.56 to 1.95 inches) thick. The entire mass was then soaked in water until the clay became soft and could be removed.

In this manner the hollow balls or bottles of caoutchouc were formed, which formerly could be found more frequently in commerce than at the present time. The cross section of the mass showed light and dark streaks which were produced according as the smoke acted on it for a shorter or longer time.

In most regions the present mode of gathering caoutchouc is as follows: The milky juice is diluted with an equal volume of water, causing the particles of caoutchouc to rise more rapidly to the surface of the mixture, from which it is gathered and separated from any adhering particles by mechanical treatment, kneading and rolling.

These manipulations are continued until a homogeneous mass is secured which is cut in disks and dried in the air. The original white color of the disks is changed to a brown, and their weight diminished by evaporation of the water.

In Central-America either alum or the juice of certain plants is added to the milky juice which has been reduced with water. It is claimed that this very much accelerates its coagulation. The caoutchouc which has been formed into a putty-like lump is lifted from the fluid, and freed from what still adheres to it in the above-mentioned manner.

To judge from the appearance of varieties of caoutchouc as found in commerce, it would seem as if different methods were used in preparing them. Many varieties contain considerable quantities of earth and sand, which have very likely become mixed with the caoutchouc by drying the milky juice in pits, while other varieties are produced in a manner similar to the one already described, namely, by forming bottles, cutting these open and pressing them into plates or kneading them into cubes, etc.

For the purpose of preventing difficulties which occur in working the crude caoutchouc, and which are especially great in working those varieties containing sand or earth, experiments have been made in skimming the cream-like layer containing the caoutchouc from the milky juice, pouring the skimmed layer into large tin cans, closing these hermetically, and transporting them to Europe, where the caoutchouc is separated with the greatest care and precaution. This method, although deserving of great commendation, has not come into general use thus far, as it very greatly increases the price of the already expensive article.

If such caoutchouc cream is repeatedly treated with acidulated water, a cohesive mass of pure caoutchouc is obtained. After the mass has been dried, it has the appearance best compared with that of light-yellow horn.

III.

COMMERCIAL CAOUTCHOUC.

It will be easily understood that a substance like caoutchouc, produced in such different localities, and derived from diverse plants, must be found in commerce in greatly varying qualities. A thorough knowledge of the different varieties is of the utmost importance to the manufacturer, but this can only be acquired by experience.

Caoutchouc is principally divided according to the locality of its production, into American, East India, Madagascar, and African. The value of a variety depends generally on its elasticity, light color, and absence of foreign substances—qualities which again depend on the treatment to which the crude milky juice may have been subjected. It is very probable that a product of uniform quality could be gained if sufficient care was used in all localities in obtaining it.

A.—American Caoutchouc.

The principal varieties come from *Carthagena, San Salvador, West Indies, Guayaquil, Guatemala,* and *Para*.

1. *Carthagena caoutchouc* comes in lumps, almost black in appearance, and weighing as much as 50 kilogrammes (110 lbs.). It is obtained in New Granada, and is very highly valued.

2. *West Indian caoutchouc.*—Numerous varieties brought into the market from the Central American States are known by this name. They come in large blocks, nearly black, and vary very much in quality. The better varieties form compact, lardaceous masses of a light yellowish-brown color (called lard caoutchouc). The poorer qualities are of a dark, nearly black color, and have a porous, spongy appearance. Guatemala caoutchouc, being more spongy than any other, is the least valued of all.

3. *Para caoutchouc* has the best reputation, which has induced several varieties derived from Brazil to be classed under this name, including some very inferior grades, carelessly prepared. Frequently, when plates or cubes of the so-called Para brand are cut apart, such large quantities of sand are found as to excite suspicion of intentional fraud in its preparation.

Para caoutchouc comes into market in the form of bottles, cubes, and plates, but the bottle-caoutchouc is being superseded by that of the plates. When a somewhat thick plate of Para caoutchouc is cut through, the outer layers generally show a dark color, while the interior of the plate is always of a light color, and frequently entirely white. From this appearance we may conjecture that the plates are dried by using heat, or obtained by cutting up larger vessels produced in the same manner as the bottles.

The lard caoutchouc comes in square cubes about 5 to 9 centimetres (1.95 to 3.51 inches) thick and 50 to 60 centimetres (19.5 to 23.4 inches) long.

B.—East Indian Caoutchouc.

There is but little difference in the principal varieties, and they are generally known by the name of a harbor they are sent from, *Bombay*, *Calcutta*, etc.

It comes into commerce in irregularly shaped cubes, formed by kneading, and contains many foreign substances. The East Indian caoutchouc is generally less valued than the American, especially the Para caoutchouc, as the American article possesses considerably more solidity and elasticity.

C.—African and Madagascar Caoutchouc.

The principal varieties are *Angola*, *Benguela*, *Congo*, and *Gaboon* caoutchouc.

Among the varieties brought from Africa, that from Madagascar brings a price equal to fine Para caoutchouc. The other varieties of African caoutchouc are of inferior quality, and known by the names of the different harbors from which they are shipped to Europe, as *Congo* caoutchouc, *Gaboon* caoutchouc, etc.

It is impossible to give a fixed rule for judging the value of caoutchouc. The article most homogeneous, least mixed with water, bark, or other impurities, has the higher value. Some buyers consider the smell as a good criterion to estimate by, but it is not a safe guide.

But, on the other hand, its elasticity is of the greatest importance, as well as its behavior while it is being worked. The greater the elasticity of the crude product, and the less mechanical treatment it requires in working, the greater is the value of the raw material.

IV.

PROPERTIES OF CAOUTCHOUC.

Caoutchouc presents many peculiarities as far as its physical and chemical properties are concerned, and a knowledge of these properties is of the utmost importance to the manufacturer. It is therefore necessary for us to bestow special attention upon the discussion of these properties, as the method employed for the mechanical treatment of the caoutchouc depends on its physical condition, and its chemical properties enable us to produce from it new substances, such as vulcanite and hard rubber.

A.—Physical Properties.

According to the method employed in producing caoutchouc, it will vary in color between a pure white and black. For instance, the outside of Para caoutchouc presents an entirely black color, which is produced by the small particles of soot adhering to it; but this variety is transparent, at least, on the edges, and thin plates of the pure article allow a reddish-brown light to pass through them.

Caoutchouc was formerly considered as possessing very minute pores, and that it was impermeable to the most subtle gases, as hydrogen. But this is an error; the great compressibility of caoutchouc alone proving very clearly the existence of many large pores, and experiments have demonstrated that gases force their way through caoutchouc plates similarly to other porous membranes.

Microscopic examinations show the pores traversing the gum in all directions, and frequently attended with branches.

The special characteristic of caoutchouc is its extraordinary elasticity, which makes it so valuable in many branches of industry. Its elasticity depends to a great extent on the temperature, decreasing in proportion as the temperature increases.

At the ordinary temperature of a room it can be stretched to a great length, but regains its original size when the tension ceases. *Gerard* observed that fibres extended six times their length could be drawn out as much more under a temperature of 100° C. (212° F.), and that the original length could be extended from 1 to 16,625, the diameter of course being decreased. Threads of remarkable fineness are obtained in this way. If an extended piece of caoutchouc is cooled off suddenly, it loses its elasticity, and does not shrink when the tension is removed, and will remain so for an indefinite period. A simpler plan of cooling the strip is by wetting it, and evaporating the water by vibrating the strip rapidly in the air. In this condition the caoutchouc resembles, but has not the rigidity of, frozen rubber. It soon regains its elasticity on being subjected to an atmosphere of 21.1° C. (70° F.), or even much below this; but rubber deprived of its latent heat by compression has been kept several weeks in an atmosphere of 26.6° C. (80° F.) without returning to its normal condition. If the heat be raised much above 26.6° C. (80° F.), or if the rubber be placed in contact with a good conductor at that temperature, it gradually recovers its latent heat, and in a few moments is restored to its original dimensions. If successive portions of the inelastic strip be pinched between the fingers it contracts in these parts, leaving the others unaffected, thereby presenting the appearance of a string of knots or beads, which may be preserved in this state for any length of time if not handled and if kept at a moderate temperature. The junctions of the different portions continue abrupt and well defined, showing that there is no tendency to distribution or equilibrium of the latent heat. When the inelastic strip is inclosed in the hand, a slight degree of coolness is felt from the rapid absorption of heat.

Considerable heat is developed in the sudden extension of caoutchouc. *Mr. Brockedon* states that he raised the temperature of an ounce of water two degrees in fifteen minutes, by collecting the heat evolved by the extension of caoutchouc thread.

PROPERTIES OF CAOUTCHOUC.

The particles of caoutchouc show great cohesion. It is very difficult to cut, and freshly cut surfaces, when pressed together, adhere as tightly to each other as before the cutting. It also adheres very tightly to other bodies, for instance, to a knife-blade. For this reason, in working it, the blades of the knives must be kept constantly wet.

We find the following records of specific gravities of different samples of caoutchouc, as well as of the same samples under different conditions:—

Best Para, taken in dilute alcohol		0.941567
Best Assam " " " . . .		0.942972
Best Singapore " " " . . .		0.936650
Best Penang " " " . . .		0.919178
Thread inelasticated by winding on a reel as above		0.948732
The same thread restored to normal condition by warming		0.925939
The above results are by Ure:		
Julian (*Dissertatio Chem. Inaug. de Res. Elas.*, &c., 1780) found sp. gr. of caoutchouc . .		0.920000
Crude caoutchouc of India . . .	Adriani.	0.966800
Black caoutchouc	Adriani.	0.945200
Prepared from juice in pure state .	Faraday.	0.925000
Determined by E. Soubeiran		0.935500
M. Payen gives		0.925000[1]

There is considerable variation in the figures given above. This might be expected from what has been stated, and will be easily explained when we consider the composition of caoutchouc.

[1] Am. Chemist, vol. ii. No. 5.

B.—Chemical Properties.

Caoutchouc is a hydrocarbon, and contains in its pure state, according to several analyses, C_4H_7, or C_6H_{10}, or C_8H_8. According to the latest examinations it is composed of $C_{45}H_{36}$. The crude caoutchouc of commerce contains oxygen. It is not a pure vegetable principle, but consists of a hydrocarbon of definite composition mixed with a small quantity of resin, the amount of which varies in different specimens. But while the percentage in its composition is of little interest to the practical man, the other substances found in some varieties are of great importance to him.

When caoutchouc is treated with alcohol certain combinations are dissolved, which are deposited in the form of crystals after the alcohol has evaporated. The crystals are easily dissolved in water, but vary very much from each other in regard to their properties. By treating them with a solution of hydriodic acid, the combinations split into new bodies, which are not fermentable varieties of sugar. The bodies soluble in alcohol have been called *Dambonite*, *Bornesite*, and *Matezite;* the first being found in African caoutchouc, the second in Borneo caoutchouc, and the last in the variety from Madagascar. Their compositions are as follows:—

Dambonite,	$C_4H_8O_3$	Dambose,	$C_3H_6O_3$
Bornesite,	$C_7H_{14}O_6$	Borneodambose,	$C_6H_{12}O_6$
Matezite,	$C_{10}H_{20}O_9$	Matezitedambose,	$C_9H_{18}O_9$

Although caoutchouc is distinguished on account of its great chemical indifference, it is very sensitive to other influences, especially to light; oxygen[1] and sulphur exert a powerful influence upon it.

If caoutchouc is exposed to the direct rays of the sun and then pressed upon a lithographic stone, the latter will take and hold printing ink where they have come in contact. Caoutchouc, which has not been exposed to the light, does not possess the property of causing this phenomenon. Therefore an exposure to the influence of light must produce a change in its constitution. But this is not yet thoroughly understood.

If caoutchouc is stored for a considerable time— several years—where it is exposed to the air, it will undergo a considerable change, at least on the surface. If such caoutchouc is treated with a solvent, with benzol, for instance, a body will be dissolved w… behind after the evaporation of the solvent, … the physical properties of a resin very rich … and not soluble in bisulphide of carbon, nor oil of turpentine (both solvents of caoutchouc). This shows

---

[1] *W. A. Miller*, in an article in the *Journal of the Chemical Society*, 1865, says: "Caoutchouc is liable to deterioration, by exposure to the action of oxygen in the presence of solar light, but it is less *rapidly* injured by their influence when in the native state than when it has been previously masticated. When subjected to the action of air, excluded from light, it does not experience any marked change, even during very long periods. It is, however, important to observe that the masticated caoutchouc is much more porous than the unmanufactured article."—*Translator.*

that, from having been for a long time in contact with the air, a partial oxidation has taken place.

C.—Behavior towards Sulphur.

The influence of sulphur upon caoutchouc is of great interest to technologists. Caoutchouc, when brought in contact with melted sulphur, absorbs it very freely, and according to the quantity of sulphur used and the temperature to which the mixture is heated, two new substances are obtained, differing very much in their respective properties.

If caoutchouc is treated with a *small* quantity of sulphur, and the mass is heated for a short time, a gray substance is obtained possessing great elasticity, which does not vary much even under changes of temperature. The product thus formed is known as *vulcanized rubber* or *vulcanite*.

But if caoutchouc is treated with a *large* quantity of sulphur at a high temperature for a considerable time, the substance gradually acquires properties so different from those of caoutchouc, that nobody could again recognize the latter in the product. It is black, possesses but little elasticity, and in regard to its physical properties may be best compared with horn. This substance is known by the name of *hard rubber*. Many manufacturers bring it into the market under the name of *keratite, hornized caoutchouc*, etc.

D.—Behavior towards Solvents.

The use of caoutchouc in solution is very important for many industrial purposes; the manufacture of waterproof tissues, of caoutchouc lacquers and varnishes depends on the use of these solutions.

Its behavior towards solvents differs from that of most substances, and in this respect very much resembles resins, with which it is generally compared. While the behavior of many substances towards solvents is such that they are either dissolved or not dissolved in them, caoutchouc, although not dissolvable by many agents, will swell up when brought in contact with them, and thereby acquire the property of being dissolvable in many substances to which, under other conditions, it is entirely indifferent.

The property of expanding in certain fluids belongs to caoutchouc in a very great degree. In many solvents, for instance, in mixtures of alcohol and bisulphide of carbon, caoutchouc swells up to thirty times its original volume, and in preparing solutions of it, this property must always be taken into consideration.

Caoutchouc cannot be dissolved in water—the latter absorbing but very small quantities of soluble substances from it—but will swell up very much when placed in water for any length of time. If a cross section of a plate of caoutchouc is examined by the microscope, it will be found that the places in the caoutchouc showing a light color, contain a large percentage of water, while the dark (outer) layers contain but very little. A thin plate of caoutchouc will assume a more uniform color in the proportion in which it dries out when exposed to the air.

Caoutchouc may absorb as much as 18 per cent. of its weight of water, and at the same time its volume increases as much as 16 per cent.

When caoutchouc is treated with absolute alcohol, its behavior is similar to that in water, except that it swells in a shorter time. But absolute alcohol cannot be considered an actual solvent of caoutchouc, as it can only absorb about 2 per cent. of it.

Among the substances capable of dissolving caoutchouc in the actual sense of the word, are the following: Ether, benzol, bisulphide of carbon, oil of turpentine, essential oils, and tar oils in general, as also caoutchene (a product gained from the dry distillation of caoutchouc). Fat oils, when strongly heated, will also dissolve it, but it is questionable whether the resulting solutions can be actually considered as solutions of unaltered caoutchouc.

The above-named substances act towards caoutchouc in such a manner that not one of them is able to hold the entire mass in solution, but can only absorb a certain percentage of it. According to experiments made by the author, a complete solution can only be obtained, *not* by using a single solvent, but by using *two* at the same time, after the material has been prepared for solution by allowing it to first swell up in one of the solvents.

For obtaining the best possible solutions, it is of the greatest importance to use the caoutchouc, as well as the solvent, in a condition as free from water as possible, and to allow the caoutchouc to swell up before treating it with the second solvent. The author of this work has made experiments as to the solubility of different varieties. The solvents used for the experiments were entirely anhydrous, and the caoutchouc was previously dried for one week over sulphuric acid. These experiments proved that different varieties are soluble in various proportions.

In the following statement we give the average figures which were the result of our experiments:—

Of 100 parts of dried caoutchouc were dissolved—

	Parts.
In bisulphide of carbon	65 to 70
" benzol	48 to 52
" oil of turpentine	50 to 52
" caoutchine	53 to 55
" ether	60 to 68

If the solutions obtained in this manner are allowed to evaporate, a colorless mass of considerable elasticity is left behind: but this mass does not possess all the properties of caoutchouc, its behavior differing from the latter, especially when subjected to heat.

The residue of the caoutchouc, after the treatment with the solvent has been several times repeated, showed a brown color similar to that of the article originally used, possessed little elasticity but considerable tenacity. When examined with a microscope immediately after it had been treated with the solvent, it appeared like a net having quite wide meshes, but these contracted considerably in drying out.

We deduce from these facts that caoutchouc consists of an insoluble base with minute pores containing the soluble parts, which resemble vegetable glue or mucein, in their capacity for distension by absorption when in contact with fluids, and facilitate the action of the solvent.

These experiments also incline us to the opinion that the elasticity of caoutchouc depends upon the quantity of soluble matter it contains.

As has been mentioned, a complete dissolution of caoutchouc requires a manipulation. We succeeded best in procuring an entirely clear solution by using the following process. The caoutchouc was allowed to distend itself by absorption of bisulphide of carbon (this absorption takes place more rapidly in a tight tank in a moderately warm place), after which 10 per cent. of absolute alcohol was added to every 100 parts of bisulphide of carbon. In a few days a complete solution was obtained, which being allowed to rest a sufficient time, all the decomposed particles were deposited on the bottom.

Treating the solution with a large amount of alcohol, the caoutchouc is deposited in a spongy form, while the foreign substances remain in solution. By pouring off the brownish solution and repeating the dissolving and precipitating, a whitish mass is obtained.

The oil of turpentine is recommended as a substitute for bisulphide of carbon—which is very objectionable on account of its great volatility and poisonous fumes—for the production of caoutchouc solutions. But the ordinary oil of turpentine as found in commerce always contains a considerable percentage of water, and does not give complete solutions. If large quantities of caoutchouc solution are to be prepared with oil of turpentine, it is advisable to free it from water by a special process.

This can be done in several ways. The simplest method is to pour the oil of turpentine with about 10 per cent. of its weight of English sulphuric acid into a well-closed tank, and allow it to stand quietly until it is to be used. The sulphuric acid forms a sediment on the bottom of the vessel from which the oil of turpentine can be easily drawn off. Instead of sulphuric acid, melted calcium chloride can be used with equal success.

If it is necessary to treat larger quantities of oil of turpentine, it is advisable to rectify it over burned lime and to pass the vapor of the oil of turpentine, before it condenses, through a nearly red-hot pipe. By this process the character of the oil of turpentine becomes greatly improved, and in consequence is much better adapted for dissolving caoutchouc.

Caoutchouc is very easily dissolved by cutting it in small pieces and placing them in boiling linseed oil, but, as has been mentioned, this solution will contain, besides the altered caoutchouc, dissolved products of it. But such a solution is very well adapted for certain purposes. When applied in a thin layer to an article and exposed to the air, it dries to a transparent mass which is distinguished by its great tenacity.

The so-called oil of caoutchouc, which is obtained by heating the latter, will dissolve the caoutchouc itself, and has therefore been recommended as a solvent, but without being especially adapted for it. For its capability of dissolving caoutchouc is only somewhat greater than that of anhydrous oil of turpentine, and the cost of producing it is comparatively great.

It seems to us that the most suitable solvents for practical purposes are bisulphide of carbon, in connection with absolute alcohol and anhydrous oil of turpentine. To be sure, benzol and coal tar oil are also good solvents for caoutchouc, but the use of these is connected with the evil, that the disagreeable odor of benzol or coal tar oil adheres for a long time and very tenaciously to the caoutchouc.

In case the work is done with bisulphide of carbon and absolute alcohol, it is advisable for the manufacturer to prepare the latter himself, and this is done in the following manner: Very highly rectified spirit of wine (containing from 95 to 96 per cent. of alcohol) is placed in a flask previously filled about one-fifth full with blue vitriol, which has been so strongly heated that the blue color has passed into white. This dephlegmated blue vitriol absorbs the last traces of water from the alcohol, and in doing this, gradually assumes its original blue color, while the alcohol standing above it has become entirely anhydrous, *i. e.*, absolute alcohol.

According to *C. Fry's* patented method, it is claimed that solutions of caoutchouc and gutta percha can be prepared with special facility if the solvent—coal tar oil or oil of turpentine—is distilled with a small quantity of caoutchouc or gutta percha. The raw oil is brought into a still, and to each 5 kilogrammes (11 lbs.) of oil, is added a quantity of caoutchouc or gutta percha amounting to between 180 and 250 grammes (6.3 and 8.75 oz.).

The solvent is distilled off, and the residue remaining in the still used for producing coarser tissues. It is claimed that solvents treated in this manner are by far the best adapted for dissolving caoutchouc and gutta percha. If they actually possess this property, they, in our opinion, acquire it simply from an admixture of decomposed products of the caoutchouc, as many of them are excellent solvents of the latter. It has been further asserted that it would be better to distil rectified spirits with the caoutchouc in the same manner as described, as the dissolving power of the solvent is supposed to be increased thereby. But since petroleum has come into use for preparing caoutchouc solutions—this subject will be discussed more fully later on—and since bisulphide of carbon and the light tar oils can be procured at very low prices, this question as to solvents for caoutchouc has lost much of that importance it formerly possessed, and at the present time it is a matter of little difficulty to prepare solutions of caoutchouc of any consistency desired.

E.—Behavior in Heat.

At a temperature of $10°$ C. ($50°$ F.) caoutchouc is comparatively solid, and not very elastic; at $36°$ C. ($96.8°$ F.) it is soft and elastic to a high degree; gradual increase of heat decreases the elasticity, and at $120°$ C. ($248°$ F.) it passes into a fluid, emitting a peculiar odor. When it has been heated to the melting point and allowed to cool off, it congeals slowly to a mass, which remains sticky for a long time. If exposed to a still greater heat, it ignites and is consumed with a bright and sooty flame.

But if treated at a high temperature in a closed tank, that is, if it is subjected to dry distillation, we obtain, besides the coal which is deposited, a quantity of gases, and a fluid called oil of caoutchouc, which, as has been stated, is a solvent for caoutchouc.

The crude oil of caoutchouc (caoutchoucine) gained by the dry distillation of caoutchouc, is a mixture of several combinations of hydrocarbons, some of which are characteristics of caoutchouc, while others make their appearance as organic substances.

Up to the present time, the following bodies have been determined as being present in the oil of caoutchouc: *Eupione, butylene, caoutchène,*[1] *isoprene, caoutchine,* and *hevéène.*[1] Caoutchène can be obtained in crystals at a very low temperature (—18° C., —0.4 F.). At —10° C. (14° F.) the crystals melt, and the fluid boils at 14.5° C. (58.5° F.). Isoprene also boils at a temperature of 37° C. (98.6 F.) and possesses the property of absorbing a large quantity of oxygen when exposed to the air, in consequence of which it is formed into a spongy, elastic mass. *Eupione, butylene,* and *isoprene* are principally found in the first part of the distillate which passes over, and must be collected in vessels thoroughly cooled off.

The larger part of caoutchouc is found in that portion of the distillate which passes over between 140° and 280° C. (284° and 536° F.). That in a pure state boils at 171° C. (339.8° F.), and, what is very remarkable, congeals only at a temperature below —40° C. (—40° F.), which is certainly a strange phenomenon for a fluid whose boiling-point is so high. Caoutchine, in common with isoprene, possesses the property of energetically absorbing oxygen.

Hevéène, which is contained in the last portions of the distillate, boils only at 315° C. (599° F.), and represents a yellow, oily fluid, having a density of 0.92.

Among the products of distillation, caoutchine and eupione are the most effective solvents.

In caoutchine, caoutchouc swells up very much, and a considerable quantity is dissolved during boiling;

---

[1] Discovered and named by *Bouchardat.—Translator.*

the solubility increases in proportion to the percentage of eupione contained in the solvent.

If oil of caoutchouc is to be prepared for the purpose of using it as a solvent for caoutchouc, the receiver in which the products of distillation are collected must be *thoroughly cooled* off, so as to retain the very volatile eupione in the distillate, and, of course, the bottles in which the oil of caoutchouc is stored must be hermetically closed.

V.

MODE OF WORKING CRUDE CAOUTCHOUC.

In describing the manner of gaining caoutchouc, and the properties of the different varieties found in commerce, we have already drawn attention to the fact that the character of the raw materials is not uniform, and that they frequently contain foreign substances, such as sand, particles of wood, etc. The first operation to which the crude material must be subjected in all cases is, to free it from all foreign admixtures; and then to change the purified caoutchouc into an entirely homogeneous mass, and only, when this is done, is it ready for further mechanical or chemical treatment.

The operations through which it has to pass may be divided according to this into two principal parts, the preparatory labor of purifying, and the treatment of the cleansed caoutchouc. While the first operation is simply a purely mechanical working of the caoutchouc, the treatment of the pure material is either a mechanical or chemical labor. When plates, threads, tubes, etc., are to be prepared from caoutchouc, it suffices simply to bring the caoutchouc into the respective forms ; but if articles of vulcanized or hard rubber are to be manufactured, the purified caoutchouc must undergo chemical treatment before the articles can be formed.

In working the crude material, it is of the greatest importance to use only caoutchouc of one and the same variety at one time, as different varieties demand different treatments.

Generally speaking, American caoutchouc is easier worked than the East Indian or African article ; if therefore American and East Indian caoutchouc, mixed together in one lot, were to be worked, the mass of the first would be refined while the other would require still further treatment.

The work is always commenced by dividing the cubes of caoutchouc into small pieces, or, still better, into shreds. These are then treated with water in order to extract the soluble parts, and the caoutchouc, when completely purified, is kneaded until an entirely homogeneous mass is formed.

The size and strength of the apparatuses used will depend on the capacity of the establishment. While formerly machines were generally used which cut the caoutchouc into irregular pieces, at the present time, they are being replaced by machines which divide it at once into fine shreds.

Such machines undergo great wear and require considerable power to work them; but the saving of time, and the complete uniformity of the product, more than repay the increased expense, and we would therefore recommend their general introduction.

As formerly mentioned, caoutchouc loses much of its elasticity when heated and becomes plastic or fusible, that is, the pieces can be easily shaped into any desired form, which they retain after being cooled off, and in this manner an almost unlimited number of pieces can be united in one.

The plates or cubes of the raw material, as they come from the factory, are placed in water, which removes the mechanically adhering impurities, and also absorbs certain substances in solution, as can be recognized by the brownish color and peculiar odor of the water.

The watered caoutchouc is then cut up. The arrangement of the older kinds of these machines resembled somewhat a straw-cutter. Sharp knife blades were set obliquely on the spokes of a wheel which revolved rapidly. The cubes of caoutchouc to be cut up were pressed by a lever against the knives, which were kept constantly wet with water to prevent the caoutchouc from adhering to them. By this apparatus the material was cut into plates or pieces preparatory for further treatment.

Although machines of the above described simple construction work very satisfactorily, they have the disadvantage of requiring frequent repairs. For instance, if the edge of the knife hits a pebble inclosed in the caoutchouc, it becomes notched and then tears rather than cuts, which makes the running of the machine more difficult, and the knives must be frequently ground. To prevent constant interruptions of the work, the knives should be so arranged that they can be easily taken off, and a number of newly ground knives kept on hand.

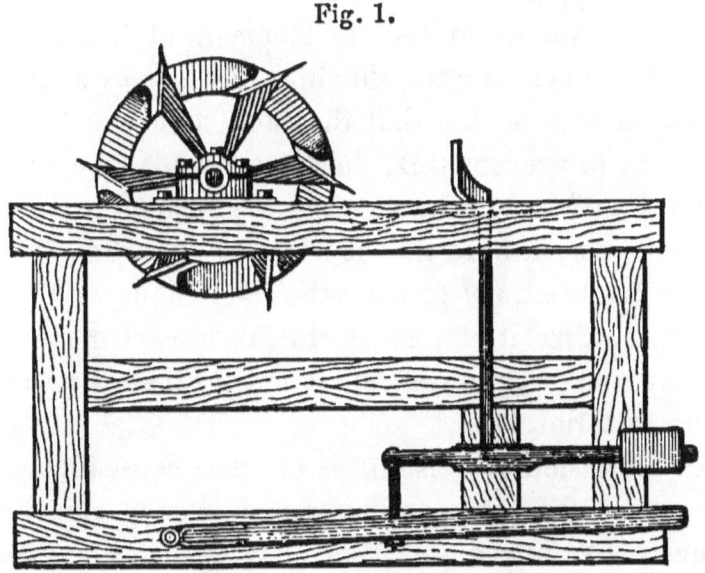

Fig. 1.

The machines for cutting up caoutchouc have been undoubtedly improved by using cylindric cutting drums (Fig. 1), the construction of which greatly resembles that of the planing machines used for working wood.

On this machine there are two circular disks connected with each other by a shaft about 20 centimetres (7.8 inches) long. The periphery of the cylinder is formed of sharp knife blades being inclined towards the superficies. The cube of caoutchouc to be cut is pressed by a lever against the cutting cylinder, and this lever is so arranged that a greater or smaller pressure can be exerted at will, an arrangement which is of especial importance when particularly soft or hard caoutchouc is to be cut. By a suitable combination of cog-wheels the cutting drum revolves at a high speed and cuts the caoutchouc, pressed against it by the lever, into shreds.

In consequence of the development of heat, which is set free in cutting the shavings, the knives or planes would become so hot that the caoutchouc would stick to them; to prevent this, the apparatus is so arranged that a jet of water falls constantly upon the knives. We will remark here that it is advisable, when cutting drums are used, *not* to treat the caoutchouc with water before cutting it up, as by this it loses a part of its solidity and is more difficult to cut up than the harder crude material.

The omission of washing is of little consequence in this case. As has been mentioned, this machine furnishes the caoutchouc in thin shavings, and one treatment of them in a hollander is sufficient to extract from them all substances soluble in water.

To obtain especially thin and tender shavings which can be cleansed in the shortest possible time, it is advisable to revolve the cutting drum as rapidly as possible, and to press the cube of caoutchouc only gently against the knives. The entire machine works very uniformly, and considerable quantities can be changed into shavings in a short time.

The shavings into which the caoutchouc has been changed by the machine are then subjected to washing. Pure water dissolves from them an average of 4 per cent. of their weight, the sand mixed with them being separated at the same time. The machines used for washing the shavings may be of various construction, but generally are provided with mechanical contrivances by which the material is always brought into contact with fresh quantities of water.

In washing materials for fabrication of paper a similar object is to be attained: the materials are to be cleansed as quickly as possible, which is done in a machine called the "*hollander.*"

The construction of the hollander allows of many variations. It is fast becoming a universal favorite, and consists generally of a round or oval trough in which water constantly circulates. In the trough are placed horizontal rollers revolving in opposite directions and at unequal velocities. The shavings thrown into the hollander are by the motion of the water constantly carried towards the rollers, which grasp, and by their unequal velocities, tear them into fine shreds as they pass through. The shreds are carried away by the current of water, and, as the form of the trough is round, are again carried towards the rollers.

In the trough are depressions into which settle the

heavy bodies from the materials in the water; these are called sand-catchers. If, for instance, caoutchouc shavings are passed through the hollander, they are torn in pieces and carried away, together with the grains of sand, by the current of water; but, while the particles of caoutchouc remain suspended, the specifically heavier substances sink to the bottom, and in this manner the cuttings can be obtained entirely free from sand, if the treatment is continued sufficiently long in the hollander.

It will be easily understood that it depends on the distance of the two rollers from each other whether the mass has to remain in the hollander for a shorter or a longer time, and whether it is to be more or less comminuted. If a very clean variety of caoutchouc is to be worked, it will require considerably less time for treatment in the hollander than if a very impure and unequal variety has to be taken in hand.

When the mass has been sufficiently treated with water in the hollander, a fluid is obtained in which swim the small pieces of caoutchouc freed from the soluble substances and mechanical admixtures. If warm water is used for treating the caoutchouc in the hollander (the water in the trough may be heated by passing steam into it), the time required for treating it may be very much shortened.

If an especially clean variety is to be worked, the treatment in the hollander may be omitted, and it is then sufficient simply to boil the mass, which has been previously torn into pieces or cut into shavings, in water, and bring them at once into the kneading-machine. At the present time the hollander has been introduced in most factories, and all varieties of caoutchouc are treated alike, that is, the crude caoutchouc is first comminuted, and then treated in the hollander until it is torn into small pieces. These are allowed quietly to collect on the surface of the water, separated from this, and then passed through the *kneading-rollers,* which must not be confounded with the *kneading-machines.*

The object of the kneading-rollers is to press the small pieces coming from the hollander into bands ready for further treatment, while that of the kneading-machines is to form compact cubes of pure caoutchouc.

Many patents have been obtained for the construction of suitable kneading-rollers, but, generally speaking, all the different constructions may be reduced to certain principal conditions. As has been stated, the object of these machines is to unite the small pieces of caoutchouc into bands. This is done by passing them through between these rollers which approach each other within certain distance; but are so constructed that they will yield in case they meet a strong resistance. This arrangement is necessary, so that, in case a mass containing pebbles should get between the rollers, it may pass through without injuring them.

This object is attained by arranging the rollers, one lying above the other in such a manner that the upper one is held in position by levers loaded with weights.

If the resistance opposed to the rollers becomes too great, the upper one rises sufficiently to allow the obstructive body to pass through, and then falls back into its former position.

The mechanism, which moves the two rollers towards each other, is so arranged that the rollers revolve with a *different* velocity, and that one of them makes, for instance, but one revolution, while the other revolves twice or three times. It will be easily understood that, on account of the different revolving velocity of the rollers, the mass which gets between them is not only *pressed*, but also *squeezed*, and that in consequence of this, a stretching of the band of rubber passing through between the rollers must take place.

The elasticity of the caoutchouc, when in a cold state, would offer considerable resistance to the rolling and pressing process, and as this elasticity disappears under a higher heat, the rollers are made hollow and heated with steam by means of connecting pipes.

In working with these machines the pieces of caoutchouc are carried to the rollers upon an inclined plane. The rollers grasp and unite them to a loosely cohering band which is passed repeatedly through between the rollers, and is finally formed into a strip of irregular widths, and although this may have many holes, it nevertheless possesses considerable solidity.

After the caoutchouc has been subjected to the above preliminary treatment, it becomes necessary to change the band into an entirely homogeneous condition, which is done by a kneading-machine.

Great improvements have been made in the construction of these machines. At this date the kneading-machines most in use consist of two rollers, obliquely inclined to each other, in a trough heated with steam by connecting pipes. The rollers, which are cut with parallel running screw threads, revolve in opposite directions. When the bands formed by the kneading-rollers are to be kneaded in this machine, the trough is heated with steam, the bands are thrown into it and the rollers set in motion. These grasp the bands and press them tightly into the screw threads, forcing out the air and water, and forming them into a compact mass.

Practical experience has shown that the homogeneity of the mass, which has been treated in the kneading-machine, increases, by storing the pieces at a higher temperature for several months before they are again worked. It seems to us not improbable that this favorable change is effected by the evaporation of the small quantity of water still adhering to the substance.

Cylindrical cubes are formed, in many manufactories, of the caoutchouc as it comes from the kneading-machines. The moulds for shaping the caoutchouc consists of cast-iron cylinders, formed of three sections fitting accurately together. One end of the cylinders is open and provided with an accurately fitting plate. When a cylinder has been filled with pieces of purified caoutchouc, the plate is placed in position and the contents of the cylinder are submitted to hydraulic pressure from six to ten days, the pressure being gradually increased. By this process the separate pieces of caoutchouc are united in a homogeneous cube which can be removed from the mould by taking the latter apart.

There are variously constructed machines used for the treatment of caoutchouc, but those are to be preferred which are provided with steam-heated rollers, as most economical of power. To obtain the caoutchouc in a homogeneous mass, it is only necessary to fold the band received at first from the rollers and re-roll it, and repeat this operation until the object is accomplished.

After the band of caoutchouc has passed several times through between the rollers, it develops sufficient heat, by the high pressure, to deprive it of almost enough elasticity to allow of further treatment, and the steam can be reduced in proportion.

When the band is sufficiently homogeneous, it is best to allow it to fall directly into the iron vessels in which it is to be pressed into a compact mass, as by doing this there will be little difficulty in completely freeing it from air-bubbles.

The longer the caoutchouc can be stored after these preparatory labors, the better will be its quality, that is, the greater will be its elasticity, tenacity, and pliability. This, in our opinion, is not caused by chemical action, but simply by the evaporation of the last traces of water still adhering to the mass.

We have stated that there are two methods of further treating the material which has been purified and formed into a homogeneous mass. It is either worked in the condition in which it is, and then the labor is purely mechanical, and, in most cases, limited to cutting the cubes into plates and threads, or this labor is preceded by a chemical manipulation which has been designated as "*vulcanizing*," although "*sulphurizing*" would be more correct. As the mechanical treatment is nearly the same for vulcanized and non-vulcanized caoutchouc, we will give a description of this labor in a later chapter of our work, and first turn our attention to the chemical treatment.

VI.

PREPARATION OF VULCANIZED CAOUTCHOUC OR VULCANITE.

We are accustomed, when hearing the word "caoutchouc," to think of a substance possessing great elasticity and extensibility. But, as has been stated, these properties exist only within certain limits of temperature; the natural heat of the human body being about the degree at which caoutchouc shows the greatest elasticity.

But if it is cooled off to a temperature of below 10° C. (50° F.) its elasticity decreases very much, and it becomes perceptibly harder. Thin plates cooled off to below zero become even breakable after being repeatedly bent backward and forward. But, on the other hand, if heated to 50° or 60° C. (122° or 140° F.), it is changed in regard to its properties in such a manner that its elasticity disappears entirely, and the caoutchouc is transformed into a tenacious mass. It is evident that the two properties just mentioned would to a very considerable extent limit the availability of caoutchouc. It would be almost impossible to make use of it in cold countries and in the tropics, and, on account of the change of the seasons, its availability would be very much decreased in our own temperate zone.

Therefore, the discovery of a process which enables us to deprive it of the property of becoming hard in the cold and soft in heat, but nevertheless retaining all its other valuable qualities, may well be designated as marking an epoch in the caoutchouc industry, and it is not saying too much, that, only since the general introduction of this process, has it become available for the manifold purposes for which it is used at the present time.

The process of which we speak has been introduced in all manufactories of rubber articles under the name of "*vulcanizing*," and consists in treating under certain conditions the purified material with sulphur.

The following are the principal properties given to it by this treatment: The preparation shows the same properties at very different degrees of temperature; if vulcanized in a proper manner, it is uniformly elastic at 20° C. below zero (—4° F.), and at a temperature higher than that of boiling water; at a lower or still higher temperature, its properties are changed in such a manner that in the cold it commences to become hard, and when exposed to heat it assumes a darker color, and finally is transformed into "*hard rubber.*"

Vulcanized caoutchouc differs very much from the crude material as far as its color is concerned, possessing a peculiar, agreeable gray tint, and has also a great many advantages over ordinary caoutchouc in regard to its chemical behavior, as it resists the action of many chemical preparations which strongly affect ordinary caoutchouc.

The process of imparting the above-mentioned properties by treating it with sulphur has been known for a long time, the first statement of the effect of sulphur upon caoutchouc dating from the year 1832, when *Lüdersdorf* drew attention to it. Whether *Goodyear* availed himself of *Lüdersdorf's* accounts, or whether he worked out the process of vulcanizing caoutchouc by self-dependent experiments, must be left undecided,[1]

---

[1] *Goodyear's* discovery of vulcanizing caoutchouc was due to close application and observation, and partly to accident. After many years of effort and disappointment, *Charles Goodyear* stood apparently as far as ever from the attainment of his object; until one day, while in earnest conversation regarding his proposed invention, he emphasized an assertion by flinging away at random a piece of caoutchouc combined with sulphur that he held in his hand. The fragment fell upon the stove, was subject to a higher heat than that to which Goodyear ever ventured designedly to subject the material, and when it was recovered it was found to possess the qualities for which he

but it is certain that the industry is indebted to him for the practical application of the vulcanizing process.

There are various methods by which the sulphur may be introduced, and these are divided into three principal groups, according to the form in which it is used:—

1. Treating caoutchouc with sulphur alone at certain temperatures.

2. Treating it with metallic combinations of sulphur, and for this purpose a large number of sulphometals have been brought into use, for instance, sulphide of potassium, trisulphide of antimony, sulphide of lead, etc.

3. Treating it with a solution of chloride of sulphur in bisulphide of carbon or refined petroleum.

Vulcanized caoutchouc prepared according to one of these methods is frequently mixed with other substances, chalk, potters' clay, zinc white, sesquioxide of iron, sand, etc. being used for this purpose. Generally the object of these admixtures is to give to the product a certain color (zinc-white, chalk, sesquioxide of iron), or, of making it rough (sand for erasing rubber). But in many cases these admixtures are

---

had so long sought; cold did not harden, and heat did not soften, the waterproof and elastic mass. And thus sprang forth the germ of an invention that has built up a new branch of manufacturing industry, given employment to thousands of operatives, and added in myriad forms to the conveniences of life.—*Translator*.

added to lessen the cost of the articles manufactured from the mass.

Properties of Vulcanized Caoutchouc.

Treating caoutchouc with sulphur changes not only its physical, but also its chemical properties. Though the chemical action in vulcanization has not yet been definitely settled, enough is known to enable us to say that an actual definite chemical combination is formed by vulcanization. The entire change in the physical conditions, but still more the behavior of caoutchouc towards solvents, denotes a distinct combination of caoutchouc with sulphur.

The elasticity of vulcanized caoutchouc does not change even within very wide limits of temperature. Freshly cut surfaces, on being pressed together, do *not* adhere, while two pieces of ordinary caoutchouc can be easily united in one piece in this manner; while ordinary caoutchouc swells up very much in benzol, bisulphide of carbon, and oil of turpentine, vulcanite does so to a very limited extent. If vulcanite is treated with ether, only the mechanically fixed sulphur is brought to the surface of the article, and is deposited there in crystals; the cross section of an article of vulcanite treated with ether is, therefore, richer in sulphur towards the exterior than in the centre. Vulcanized caoutchouc when placed in water absorbs less of it than non-vulcanized caoutchouc, and, generally speaking, fluids do not penetrate the former so easily as the latter. But towards gases the behavior of both varieties of caoutchouc is about the same, vulcanite, for instance, allowing of the passage of considerable quantities of illuminating gas.

The peculiar constitution of vulcanite is shown most plainly in treating the material with solvents. If bisulphide of carbon, or ether, is allowed to act upon vulcanized caoutchouc, about 4 per cent. of unchanged caoutchouc is dissolved as well as the free sulphur, which may be present. If vulcanized caoutchouc is treated for months with a compound of 10 parts of bisulphide of carbon, and 4 parts of absolute alcohol, 35 per cent. of the entire mass can, according to experiments made by *Payen*, be brought into solution. 10 per cent. of this is unaltered sulphur; the other 25 per cent. is easily soluble, and consists very likely of a combination of caoutchouc with sulphur, which is not thus far thoroughly understood.

Experiments made by *Payen* and others have demonstrated that 1 to 2 per cent. of the weight of the caoutchouc originally used, is sufficient for forming vulcanite, but far more sulphur is added in practice, and a great quantity of sulphur in a free state is present in the vulcanite as a mechanical admixture. But the excess of sulphur mixed with the vulcanite does not remain chemically indifferent; when the vulcanite is stored for some time, the sulphur combines with it, and changes it, so that in regard to its properties it approaches hard rubber, which is very rich in sulphur.

Articles prepared from vulcanite containing a great

excess of sulphur entirely lose their elasticity when they are stored for a few years, become as hard as wood, and break very easily, for instance, tubes, if they are bent backward and forward several times. Articles of soft rubber having become useless can be worked into hard rubber.

If vulcanized caoutchouc is boiled in fluids possessing the power of absorbing sulphur, the excess of sulphur is dissolved, and the caoutchouc contains only the chemically fixed sulphur. Vulcanite treated in this manner is known as *desulphurized* rubber, and resembles in appearance the ordinary caoutchouc; but has retained its indifference towards changes of temperature and chemical agents, and must be considered as the most valuable variety of vulcanized caoutchouc.

Vulcanized caoutchouc is prepared according to various methods, but only a few of these methods have been adopted in practice, as the others, although they appear all right theoretically, do not yield a product of a corresponding quality.

VII.

VULCANIZING WITH PURE SULPHUR.

Although caoutchouc will absorb sulphur merely melted—sulphur melts at 113° C. (235.4° F.)—it will require a long time before the operation is finished. For the purpose of hastening the labor, it will become necessary to increase the temperature from 150° to 170° C. (302° to 338° F.), and keep it there for two hours. But the greatest care must be exercised not to go much above this temperature, as the caoutchouc would then acquire the properties of hard rubber, and not those of vulcanized caoutchouc or soft rubber.

Although at the present time the method of vulcanizing by plunging into melted sulphur has been almost abandoned, yet, on account of its simplicity, we have not omitted to subject it to an examination. Our experiments have shown that, as sulphur penetrates but slowly the interior of caoutchouc, it is scarcely possible to obtain in this manner a uniform product, the outer layers of the caoutchouc being generally more than vulcanized (*i. e.*, they commence. to approach hard rubber as far as their properties are concerned), while the interior strata show the right condition—but the core of the caoutchouc remains entirely unchanged.

To obtain an entirely uniform product by this method—and only such answers all demands—it is absolutely necessary to subject the caoutchouc to a vigorous, mechanical treatment, while it is in contact with the sulphur. Experiments have been made to facilitate and accelerate the absorption of sulphur with the assistance of superheated steam. But in no case were the results of this labor satisfactory—it was not possible to obtain an entirely uniform product.

The part of the labor of vulcanization where the mass is exposed to a higher temperature, is technically designated as "*burning*," and it is absolutely necessary that the laborer who is entrusted with this work should be very careful and experienced. Even if the greatest care be observed, the labor of uniformly vulcanizing larger quantities of caoutchouc by this method is not always successful. We are of the opinion that this is to be attributed, not so much to any difficulty the sulphur may find in penetrating the caoutchouc, but rather to the fact that the latter is a bad conductor of heat; it being, as is well known, such a bad conductor of heat that the end of a very small piece can be held between the fingers while the other end is being heated to combustion.

When caoutchouc is plunged into hot sulphur, and the mass is then heated, it will be observed that, although it absorbs the sulphur, it does *not form a chemical combination with it.* When the mass, which may appear entirely homogeneous to the naked eye, is examined with the microscope, the particles of coautchouc and those of sulphur can be distinctly distinguished from each other. The chemical combination of both substances does not take place until the mass is burned, *i. e.*, is heated to above $150°$ C. ($302°$ F.), when vulcanized caoutchouc is formed. Taking this circumstance into consideration, a process has been adopted, of which the following are the principal features.

A.—Mechanical Combination of Caoutchouc with Sulphur.

Caoutchouc and sulphur are first intimately mixed together by mechanical means, the mixture is then heated or burned to a temperature between 150° and 170° C. (302° and 338° F.), the chemical combination of both substances taking place at this degree of heat. If the temperature required for burning is accurately regulated, a product is obtained which is entirely homogeneous and answers all demands.

If it is desired to prepare vulcanized caoutchouc according to this method, which will be shortly described more fully, the mass of caoutchouc which has been obtained after treating the crude article in the hollander, can be used at once without making the mass entirely homogeneous by passing it repeatedly through the kneading rollers, or without forming the bands into cubes. The work is simplified by performing the labor of making the mass homogeneous while mixing the sulphur with the caoutchouc.

It has already been mentioned that larger quantities of sulphur are used in the vulcanizing process than are actually necessary to produce vulcanite; according to theory, 1 to 2 per cent. of the weight of the crude caoutchouc of sulphur is sufficient, but in practice a quantity of sulphur amounting to from 12 to 24 per cent., and still more of the weight of the caoutchouc is taken.

The sulphur may be used either in the form of finely powdered roll-sulphur, or flowers of sulphur, the latter offering the advantage of being already in the form of a fine powder, but sometimes considerable quantities of sulphurous acid adhere to them, which possibly might exert a bad influence upon the qualities of the vulcanite. If flowers of sulphur are to be used, it is absolutely necessary to remove the last trace of sulphurous acid by washing them with water, and then to dry them completely.

The labor is commenced by passing the mass of caoutchouc, which comes from the hollander and consists of very small pieces, through the kneading rollers, so as to form them into a loose band; the rollers used for this purpose should be such as can be heated by steam. The loose band obtained in this manner is passed a second time between the rollers, and at the same time is powdered with sulphur. When all the sulphur has been put in the band, it is advisable to go systematically to work to knead the sulphur into the caoutchouc. The best method of doing this work is to double the band together, and to repeat the stretching until a mass is obtained which, when examined by the naked eye, appears to be entirely homogeneous.

When this is the case, the mass, as far as its properties are concerned, is the same as pure caoutchouc which has been passed through the kneading rollers. It only represents a *mechanical* compound of both substances, but a *chemical* combination has not been formed. The mass is elastic in the cold, of a brownish color, nearly unelastic in heat; freshly cut surfaces adhere tightly together, solvents of sulphur easily dissolve the *entire quantity* of the sulphur which has been kneaded in.

As freshly cut surfaces of this mass will adhere together, this property is made use of, and articles to be vulcanized are formed from it before undergoing the burning process. (As has been mentioned, freshly cut surfaces of *vulcanite* will *not* adhere together.) To change the mass into vulcanite it is subjected to burning, and this can be done with the bands, and the cubes prepared from the bands; but most frequently the articles themselves, which have been formed from the mechanical compound of caoutchouc and sulphur, are subjected to the burning process, as it is very difficult to manufacture them from the completely vulcanized caoutchouc by cementing them together, etc.

B.—Burning of the Caoutchouc and Sulphur Mass.

As has been said, the chemical combination between caoutchouc and sulphur takes place only when the compound of both bodies is heated to a certain temperature. Manufacturers disagree as to the degree of heat which is required for burning, and the reason for this difference of opinion is, that the various varieties of caoutchouc do not behave in the same manner. All Asiatic varieties (from the East Indies, Java,

Borneo) requiring less time for vulcanization than the finer American.

The time required for vulcanizing depends on the thickness of the article to be treated; caoutchouc is a bad conductor of heat, and, in consequence of this, much more time is required to thoroughly change thick articles into vulcanite than very thin plates. If a mass is not heated enough, or for too short a time, its surface alone will be changed into vulcanite, while the interior of the mass remains unaltered.

The degrees of heat between 120° and 150° C. (248° and 302° F.) may be considered as the limits of temperature in which vulcanization will progress in a correct manner. But in some cases it is allowable to go above this limit, and to raise the temperature to 170° C. (338° F.), but this increased temperature must be used for a very short time only, as, if continued too long, the articles will lose their elasticity, at least on the surface, and become hard, or, if burnt too strongly, will, in a short time, show some brittleness.

According to special experiments we have made as to the temperatures, which are absolutely necessary for the correct practice of the vulcanizing process, we have come to the conclusion, that it is a principal condition to heat the mass above the melting-point of sulphur. Now, as sulphur melts at 113° C. (235.4° F.), theoretically it would be sufficient to heat the mass, to be burned, to a temperature only somewhat above 113° C. (235.4° F.), for the purpose of carrying on the vulcanizing process.

Our experiments have also shown this to be a fact, that it is possible to change caoutchouc, which has been very intimately compounded with sulphur, into vulcanite by heating it to 115° C. (239° F.), but this fact is of very little, or actually, of no value whatever, for practical purposes, as the time, which seems to be required for the progress of vulcanization at such a low temperature, is so long that the manufacturer cannot afford to do it. Therefore it depends entirely on the conditions under which the manufacturer works, what temperature he is to use in vulcanizing. If he consumes American caoutchouc, and has to vulcanize articles with thick walls, it is absolutely necessary for him to have a higher temperature than is required for East Indian material, and articles with thin sides, perhaps not more than 1 centimetre (0.393 inch) thick.

It would, therefore, be advisable, for the above-mentioned reasons, not only to work one and the same variety of caoutchouc at one time, but also to make it a point to subject only articles which have no great difference in their thickness to the burning process at one time.

The comparatively narrow limits of temperature in which the burning process must be carried on, and, caoutchouc being a bad conductor of heat, it becomes necessary to use special apparatuses in this process. These apparatuses must be so constructed, that, even in a large room, the heat can be so regulated as not to vary more than 2 or 3 degrees during the entire time required for the work.

Various means have been proposed for attaining this object, but not all of them can be regarded as suitable. Some, for instance, have recommended the use of alloys which melt at a certain temperature. It is true, there are alloys which melt at a temperature far below that of boiling water, as, for instance, *Wood's* metal compound and *Rose's* alloy. These alloys are composed of lead, tin, bismuth, and cadmium, and their melting-points can be changed at will by a corresponding change in the quantities of cadmium and bismuth.

Independent of the fact that, on account of the high prices of cadmium and tin, the preparation of these alloys would be a very expensive business, and further, that it is very difficult to work with a metal bath, there are still other reasons why these metal baths are not available for practical purposes. No doubt it is true that freshly prepared alloys melt at a certain temperature, but it is also true that their melting-points change when they are frequently heated. Namely, when these masses of metal are frequently melted, alloys of definite composition are formed whose melting points lie far higher than those of the compounds originally used.

These evils are of sufficient importance to forbid the use of metal baths for vulcanizing caoutchouc in manufactories. The use of oil baths or highly concentrated solutions of salt would be preferable.

The technology of heating has advanced so far at the present time that no complicated apparatuses are required to obtain a very uniform temperature, even in a large room, and heated air or superheated steam is now generally used for the purpose.

If heated air is to be used, small chambers of brickwork are built; these are provided with air-tight doors and the floor consists of well-joined iron plates. Several of such chambers adjoin each other, and are heated by one fire. The flues are arranged in such a manner that the gases of the fire must pass frequently to and fro, under the iron plates of the bottom of the chambers, and heat the air in them very uniformly.

The articles of caoutchouc to be burned are put upon a frame placed close to the ceiling of the chambers, and the fire is so regulated that the thermometer, which is placed behind glass plates fitted into the doors of the chambers, shows as uniform a temperature as possible (130° to 140° C., 266° to 284° F.). If the work is done on a small scale, *Meidinger's* furnace may be used for heating purposes. The supply of air to control combustion, and consequently the temperature of the air passing upwards, is regulated by opening or shutting the slide on this furnace.

For larger manufactories, which should be provided with steam-power for driving the machines used for treating the caoutchouc, it is more suitable to perform the burning by means of superheated steam. The boiler used for this purpose must be made strong enough to bear a pressure corresponding to a heat raised to 150° C. (302° F.). When superheated steam is used, the temperature can be so kept that it will vary only a few degrees in the prescribed limits, by regulating the admission of steam.

C.—Burning Apparatuses.

The arrangement of apparatuses used for burning the vulcanite by means of superheated steam varies very much. A low box of wood may be used, on the bottom of which is placed a serpentine pipe, through which the superheated steam passes, or the steam is allowed to circulate between the sides.

But the most suitable apparatus for vulcanizing articles of caoutchouc is one constructed of iron plates

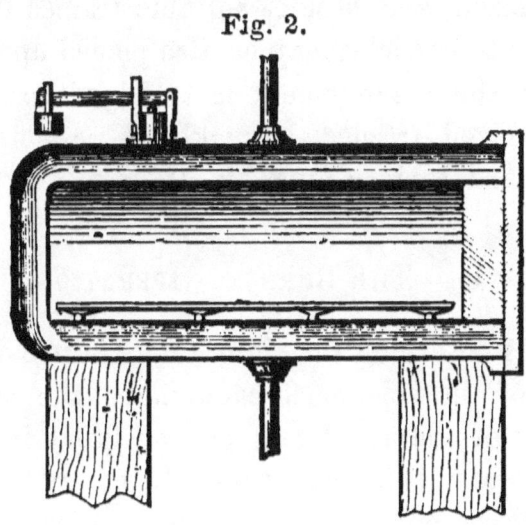

Fig. 2.

riveted together, and somewhat resembling a steam boiler. The accompanying illustration (Fig. 2) represents the arrangement of such an apparatus. It consists of an iron chamber, nearly square, and closed on one end. This chamber is surrounded by a second in such a manner as to leave a distance of about 10 centimetres (3.39 inches) between the two chambers. On the upper side of the exterior chamber are placed the steam-feeding pipe, a safety-valve, and a thermometer; on the lower side is a pipe which carries off the condensed water. The outer chamber is covered with non-conductors of heat (wool, wood, cut straw, etc.) to prevent a loss of heat.

The open end of the inner chamber can be closed by a close-fitting door. On the bottom of this chamber are placed rails upon which run the wheels of a small iron carriage containing the articles of caoutchouc to be vulcanized, and besides suitable frames for the reception of the articles may be also placed upon it. The object of this arrangement is to allow the removal of the vulcanized articles as quickly as possible from the chamber and replace them by others to be vulcanized.

D.—The Burning Operation.

Besides observing the right temperature during the burning of the vulcanized caoutchouc, the mechanical treatment of the articles is also of great importance. That is to say, it is not practicable simply to place the articles to be burned in the burning apparatus, as in the high temperature to which they must be exposed they would soften so much as to entirely lose their shape. Therefore all articles with a surface of a definite curvature must be placed upon wooden or metal forms before placing them in the chambers.

To prevent the caoutchouc articles from adhering to the forms, it is necessary to coat the latter with powdered talc, a mineral of a soapy feel. The talc is put in a linen bag and the forms, as well as the articles, are dusted with it. Small curved articles may be placed directly in tin boxes filled with powdered talc and so subjected to the burning process.

To prevent thick caoutchouc plates from warping during the burning process, they are pressed between iron plates. Thin plates of vulcanite are first shaped from the vulcanized caoutchouc, then completely dusted with powdered talc and covered with a linen cloth. In this covered condition they are wound upon a wooden roller, forming a cylinder of the length of the plate and of course varying in diameter.

If the length of the cylinder is in a great ratio to its diameter, the time of burning is prolonged, as it will require some time to heat the mass to the burning temperature.

To avoid losing time the articles should be sorted that are to be burned. The thin and thick should have separate treatments as the latter require more time to vulcanize than the former.

Many manufacturers perform the burning in two separate operations: In the first the apparatus is heated only for a short time at not over 140° C. (284° F.). The articles that were too soft to be taken from the moulds attain by this operation sufficient tenacity to be burned by themselves. It is evident that this plan is attended with loss of time and heat, and that it is more advantageous to finish the work in one operation.

Vulcanization is modified to suit such cases as require thoroughly and uniformly vulcanized caoutchouc. The articles are made in the usual manner of pure caoutchouc by heating sufficiently to destroy its elasticity, and are then pressed into shape. They are now coated with a saturated solution of sulphur in bisulphide of carbon. This coating should be as uniform as possible over the entire surface.

The following method suffices for articles needing only light treatment. After their formation in pure caoutchouc, the articles are immersed in oil of turpentine and allowed to remain until their surfaces have become somewhat sticky, when they are dusted with finely powdered sulphur, care being taken to avoid an excess of it.

The articles treated either with the solution of sulphur or simply dusted with sulphur, are burned in the same manner as those prepared from caoutchouc compounded with sulphur. By this method the caoutchouc may also be changed into vulcanite, provided the articles have been prepared with sufficient care, but in regard to thicker articles it is impossible to produce a homogeneous product in this manner; a cross section of the pieces will show plainly that vulcanization has taken place on the surface only, while the interior parts of the caoutchouc remain unchanged.

VIII.

VULCANIZING WITH THE ASSISTANCE OF CHLORIDE OF SULPHUR.[1]

Although the labor required for the preparation of vulcanite by the above described method cannot be called an excessively heavy one, it cannot be denied that this method possesses also some objections. These may be found in the fact that the mass of caoutchouc must repeatedly pass between the rollers before it can be entirely homogeneously combined with the sulphur, and further that a special apparatus is required for burning, and that this operation demands great attention and care from the workmen.

But all the above-named evils are entirely dispensed

---

[1] This is known as *Parkes'* cold vulcanizing process and was discovered and patented by *Alexander Parkes*, of Birmingham, England. The following process devised by *Gaulthier de Caulbry*, is claimed also to effect the vulcanizing of caoutchouc at common temperatures, although it has never obtained a strong foothold in practice. If an intimate mixture is made of flowers of sulphur and dry chloride of lime, a decided odor of chloride of sulphur will be noticeable while simultaneously the temperature of the mixture is appreciably elevated and the mass becomes plastic by the softening of the sulphur. If a mixture of this kind in which the sulphur is in great excess is added to caoutchouc softened in bisulphide of carbon, it effects the change called vulcanization at an ordinary temperature or upon slight warming. When the mixture contains an excess of the chloride of lime, the mass does not become pasty but remains pulverulent.—*Translator*.

with when the work of vulcanizing is done with a solution of chloride of sulphur. But chloride of sulphur should never be used by itself, as the pure article exerts too energetic an effect upon the caoutchouc, and for this reason a strongly reduced solution of it should be used in all cases.

The work can be done at an ordinary temperature, and the manufacturer has it in his power to proceed with the vulcanization to any desired extent. But only *thin* articles can be changed into vulcanite by this method. If, for instance, plates having a diameter of a few centimetres are to be changed, it is best to employ the method of incorporating the sulphur with the caoutchouc, and subject the compound to the burning process.

Chloride of sulphur, the preparation and properties of which will be given further on, possesses the faculty of quickly penetrating caoutchouc and changing it into vulcanite. The articles prepared from caoutchouc are placed in a weak solution of it, and allowed to remain for a sufficient time, when they are removed and thrown into warm water, or dried in a current of warm air.

Either bisulphide of carbon, or, latterly, completely anhydrous petroleum, refined in a particular manner (the process will be described further on), is used as a solvent for chloride of sulphur. Manufacturers recommend solutions of different degrees of concentration, according as thicker or thinner articles are to be vulcanized.

The following has been recommended for thinner articles: Chloride of sulphur, 1 part by weight; bisulphide of carbon, 30 to 40 parts by weight; and plunging the articles into this solution for 60 to 80 seconds.

For thicker articles it has been recommended to use a solution of chloride of sulphur 1 part by weight; bisulphide of carbon 60 to 80 parts by weight, and to allow the articles to remain in the solution for three, four, to five minutes. Articles of extra thickness must be repeatedly plunged into the fluid until vulcanization is completed, when they are washed and dried.

Our experience is that it is more suitable in all cases to work with a reduced solution of chloride of sulphur, as it is then possible to stop vulcanization at a definite moment. If the articles are allowed to remain too long in the solution, over-vulcanization takes place very easily, that is, the surface of the articles becomes hard and brittle.

Solutions of chloride of sulphur, whether prepared with bisulphide of carbon or with petroleum, exert an injurious effect upon the health of the workmen. This is especially the case with bisulphide of carbon, as it emits its poisonous vapors in addition to those from the chloride of sulphur, making the operation of vulcanizing exceedingly disagreeable.

Precautionary measures should be taken to protect the workmen from the injurious effects of the vulcanizing solution. The solution should be contained in plate-glass tanks, with lids removable horizontally by a lever worked by the foot of the operator, and kept in position by weights when freed from the action of the lever.

If the articles are to be dried after removal from the vulcanizing fluid, it is best to place them in a box through which a current of air at a temperature from 30° to 40° C. (86° to 104° F.) is forced by a ventilator. To prevent the vapors of the bisulphide of carbon from annoying the neighborhood it is advisable that the mouth of the pipe which carries them off should be below the grate of the furnace, where they come in contact with the fuel, and are burned into carbonic acid and oxides of sulphur.

The drying must be done very quickly, to prevent the solution of chloride of sulphur still adhering to the surface of the articles from exerting a further effect upon the caoutchouc, as this might cause over-vulcanization, that is, the caoutchouc might become hard.

The appearance of the annoying vapors, as well as over-vulcanization, can be avoided by throwing the articles, taken from the solution, into a vessel filled with warm water. When chloride of sulphur comes in contact with water, it is at once decomposed into hydrochloric and sulphurous acid, and, in consequence of this, the progress of vulcanization ceases the moment the articles are thrown into the water.

Small articles, especially those of sharp forms produced by stamping, can be perfectly vulcanized by this method, and their contour preserved, which they would lose by melting during the burning process, if prepared by other methods.

When larger articles are to be vulcanized with the aid of chloride of sulphur, they should remain for a longer time in the solution, and may be placed back into it, if, after being washed off, they appear to be imperfectly vulcanized.

Articles of special beauty should be thoroughly washed after the vulcanizing process, and then placed in caustic soda of moderate strength at the boiling point, and allowed to remain in it from 50 to 70 minutes. The caustic soda dissolves the free sulphur, and the surface of the article will present a uniformly gray color.

It is immaterial whether solutions of chloride of sulphur in bisulphide of carbon, or solutions prepared with the aid of anhydrous petroleum are used in carrying on this work. The latter method can be decidedly recommended as being much cheaper than with bisulphide of carbon, and less injurious to health.

Preparation of Chloride of Sulphur.

The chloride of sulphur used for preparing the vulcanite can be manufactured in the caoutchouc factory, by conducting *entirely dry* chlorine gas over *completely dried* powdered sulphur. The sulphur, which must be dried immediately before the operation, is placed in a tubulated retort, which is provided with a thoroughly cooled off condenser. A pipe is placed in the tubulure through which the chlorine passes. To obtain the chlorine in an entirely dry condition, it is conducted through a pipe filled with pieces of pumice stone, which have been saturated with sulphuric acid. It is an absolute necessity for the success of the operation that *both* materials should be *entirely dry*, as chloride of sulphur becomes decomposed as soon as it comes in contact with water.

The action of both bodies commences as soon as the retort containing the sulphur is heated. A reddish-yellow fluid, consisting of a solution of free sulphur with chloride of sulphur, is collected in the receiver. To free the latter from the sulphur, the fluid is distilled until it *boils exactly at* 130° C. (266° F.), but this distillation of the fluid may be omitted if the presence of free sulphur does not exert a disturbing influence upon the vulcanizing process.

Pure chloride of sulphur is a fluid, of an orange-yellow color, of great density (1.680), fumes strongly when exposed to the air, and throws off vapors of hydrochloride. Brought in contact with water, chloride of sulphur breaks up very quickly into hydrochloric acid, sulphur, and sulphurous acid.

As chloride of sulphur severely attacks the mucous membranes of the nose and mouth, as well as the eyes, and produces a convulsive cough and difficulty of breathing, followed, if the air impregnated with its poisonous vapors is inhaled for any length of time, by lung and throat diseases, the greatest precaution should be observed to protect the workmen from its evil effects. It should be kept in bottles with well ground stoppers.

Refining of Petroleum.

To avoid the deleterious effects on the health of workmen by bisulphide of carbon, as described under that head, it is desirable to substitute some equally effective solvent free from such serious objections. Petroleum is well adapted for this purpose, if it is absolutely free from water, as otherwise the chloride of sulphur would be decomposed.

To make petroleum anhydrous, it is placed in a vat lined with lead, and provided with a stirring apparatus. It is mixed with the tenth part of its weight of English sulphuric acid, and after the mixture has been thoroughly stirred for several hours, it is allowed to rest quietly. The petroleum, which swims on the top, is then brought into a still, and to every 100 parts of it one-quarter part of burned lime, finely powdered, is added to fix the last traces of acid. The petroleum is then distilled off. To prevent the rectified petroleum from absorbing water it should be kept in well-closed glass bottles (the so-called balloons used for storing hydrochloric acid are especially well adapted for the purpose).

IX.

USE OF SULPHO-METALS FOR VULCANIZING.

Sulpho-metals may also be used for vulcanizing instead of sulphur, or chloride of sulphur. The sulphides of the alkali-metals (sulphur, potassium, sodium, calcium, barium) are especially well adapted for the purpose, and, for this reason, the product, vulcanized by these means, has also been called "*alkalized caoutchouc.*" This designation is decidedly incorrect, as not the "*alkali*," but the "*sulphur*," effects the vulcanization. Besides the above-mentioned materials, many combinations of heavy metals with sulphur, such as sulphides of antimony, lead, bismuth, etc., are also adapted for the preparation of vulcanite, but are seldom used for the purpose.

Calcium sulphide can be obtained by forming finely powdered gypsum into a dough with charcoal, adding to this a little oil and subjecting the paste to a white heat; barium sulphide may be prepared from heavy barytes in the same manner.

The sulphides, changed into a fine powder, are compounded with the caoutchouc in a similar manner as sulphur, and the articles formed from the mass are subjected to the burning process.

Antimony trisulphide and lead sulphide, as well as bismuth trisulphide (the use of this would be rather expensive) have also been recommended for vulcan-

USE OF SULPHO-METALS FOR VULCANIZING. 95

izing caoutchouc in a similar manner as by means of the above named sulpho-metals. It was claimed that vulcanized caoutchouc, prepared by this method, possessed a far greater degree of solidity, than that prepared with sulphur alone, but this has not been confirmed by practical experiments. As the manufacture or preparation of the sulpho-metals is always connected with considerably more expense than the use of pure sulphur, the various methods which have been proposed have never gained a foothold in the practical industry, and we therefore content ourselves with having alluded to them.

But one of these methods is an exception, namely that of using a concentrated solution of liver of sulphur (penta-sulphide of potassium) for vulcanizing. The solution is obtained by melting carbonate of potassium together with sulphur. By using a smaller quantity of sulphur, trisulphide of potassium is obtained, with more sulphur, penta-sulphide of potassium.

The following quantities are used for preparing the last-named combination:—

	Parts.
Carbonate of potassium	276.8
Powdered sulphur	256.0

But as the above mentioned figures represent pure carbonate of potassium (*i. e.*, 100 per cent. potash), and the commercial article is never entirely pure, a smaller quantity of sulphur corresponding to the percentage of potash in the carbonate must be taken.

The substances are first reduced to a fine powder and dried, then quickly mixed together—as potash absorbs a great deal of moisture from the atmosphere—and fused in a crucible in quantities of from 20 to 25 kilogrammes (44 to 55 lbs.). The crucible must be comparatively large, as the mass, in consequence of the escape of carbonic acid, strongly effervesces during fusing. The mass is kept at the fusing point until all effervescence has ceased, and is then scooped with an iron ladle into flat mould-like vessels of sheet iron, in which it is allowed to congeal. The congealed mass of penta-sulphide of potassium (on account of its brown color it is also called *liver* of *sulphur*) must at once be put in well-closed glass vessels as it becomes decomposed if exposed to the air.

The fluid for vulcanizing purposes is prepared by making a concentrated solution of the penta-sulphide of potassium; it should show 25° Baumé. This is put in a porcelain vessel and quickly brought to the boiling point. The articles to be vulcanized are then immersed in the fluid and allowed to remain there until vulcanization is complete. This process has been recommended by *Gerard*, and should it prove itself available for all purposes, it seems to us, that eventually it will supersede all other methods of vulcanizing, as it has the advantage of being entirely innoxious and at the same time inexpensive. But to judge from the experiments we have made it would seem as if it required some further improvement, as we only succeeded in changing thin pieces into vulcanite, thicker pieces not being uniformly vulcanized.

Vulcanite Masses.

It is absolutely necessary to use pure vulcanite for manufacturing articles which are to possess in a very high degree the property of elasticity combined with tenacity. By *pure vulcanite* we mean a substance which has been prepared from *pure caoutchouc* and *sulphur* according to one of the methods described.

In some cases elasticity is not so much required as cheapness of production, as toys for children, little tubs, cups and saucers, etc. Such articles contain a very small percentage of caoutchouc, but many admixtures.

The substances used in the manufacture of an article depend on the properties it is to possess; if articles of a light color and little weight are desired either fine white pipe-clay, chalk, or magnesia are mixed with the mass to be vulcanized. For white masses of greater weight, oxide of zinc is used, or sulphate of lead which, being a waste product of chemical factories, can be procured at a comparatively small cost.

Cinnabar, round lake, sesqui-oxide of iron (caput mortuum, colcothar) are generally added to produce a red color; ultra-marine, or smalt may be used for blue; chrome yellow furnishes the yellow color; a mixture of chrome yellow and ultra-marine, green; colcothar and ultra-marine, violet, etc.

A uniform coloring of the caoutchouc can only be effected by carefully kneading the coloring substances into the mass. But it can also be completely colored by producing certain chemical combinations in the mass itself. As this process has not yet been introduced into general practice, we give in the following a few receipts for coloring caoutchouc as well as gutta percha.

For *black*, a fluid is used consisting of:—

Blue vitriol	1 kilogramme		(2.2 lbs.)
Water	10	"	(22 lbs.)
Caustic ammonia	1	"	(2.2 lbs.)
Muriate of ammonia	0.5	"	(1.1 lb.)

The blue vitriol is dissolved in water together with the muriate of ammonia and the caustic ammonia is then added.

For *green* the following are used:—

Blue vitriol	0.5 kilogramme		(1.1 lbs.)
Muriate of ammonia	1	"	(2.2 lbs.)
Burned lime	2	"	(4.4 lbs.)
Water	10	"	(22 lbs.)

For *violet*:—

Blue vitriol	0.25 kilogramme		(0.55 lbs.)
Sulphate of potassium	1	"	(2.2 lbs.)
Phenicine	0.25	"	(0.55 lbs.)
Water	10	"	(22 lbs.)

The articles to be dyed are boiled in their respective fluids from 15 to 30 minutes, but articles somewhat thicker must be boiled for a longer time to make the coloring uniform. The dyed articles can then be vulcanized in the usual manner.

Masses of caoutchouc which require to be rough are compounded either with powdered pumice-stone or the finest drift sand. The last two are added to the mass at the time when the sulphur is kneaded in, or in case the process of vulcanizing with a fluid (chloride of sulphur) is used, they are added to the pure caoutchouc. To insure a completely uniform mixing of the added ingredients, as clay, magnesia, pumice-stone, etc., it is absolutely necessary that they should be powdered as fine as possible, carefully washed and then thoroughly dried, as the presence of moisture is very detrimental.

White or black pitch is also added to the caoutchouc mass for the purpose of manufacturing cheap products; many manufacturers assert that the properties of the vulcanite are essentially improved by such additions, but this assertion is not true, and is probably only made for the purpose of hiding the real object of these admixtures, namely, the production of masses in a cheap manner. In the following we give some receipts for preparing vulcanite mixtures. These receipts have been tried and found useful.

White Caoutchouc masses.

	Parts by weight.
Caoutchouc	100
Sulphur	10 to 20
Chalk	40 to 60
Magnesia	5 to 40
Oxide of zinc	20 to 30

If the composition is to be colored, the coloring agent takes the place of one of the other components, either of chalk, magnesia, or oxide of zinc. This compound endures burning at a high temperature and can be finished in one operation.

Cheap Caoutchouc masses with an addition of Resin.

	Parts by weight.		
Caoutchouc	100	200	200
Sulphur	25	25	50
White pitch, or	15	18	25
Pine resin	12	60	20

Masses prepared with an addition of pitch or resin cannot stand a high burning temperature. When heated to 140° or 150° C. (284° or 302° F.) they melt so much that the mass would collapse over the mould. Therefore in burning the caoutchouc in this case it should only be heated very little above the melting point of sulphur (113° C., 235.4° F.) A temperature between 115° and 120° C. (239° and 248° F.) will be most suitable for the purpose. Thinner articles prepared from these masses possess a great degree of elasticity but thicker pieces show a considerable decrease of it.

Deodorizing Vulcanite.

All articles manufactured from vulcanized caoutchouc possess a disagreeable odor, perceptible even after the articles have been in use for months. As this odor is very repugnant to some persons, and we know from experience that many will not use articles of vulcanite simply for this reason, it becomes a matter of importance to remove this objection, especially from articles intended for personal use (pocket-books, cigar cases, etc.).

This is done in various ways: the articles are either exposed to a constant high temperature, or they are treated with animal charcoal. It is true, heating alone effects a decrease of the odor, but to get rid of it entirely the heating must be continued for many days, and this method is, therefore, not available in practice. Animal charcoal possesses in a high degree the property of absorbing odors. For our purpose it is best to use it in the form of powder.

A large number of articles can be deodorized at one time, by treating them in the following manner: Cover the bottom of a sheet-iron box about 2 centimetres (0.78 inch) deep with animal charcoal, upon which place the articles. Fill in the spaces between them, and cover with a top layer about 2 centimetres

(0.78 inch) thick. Upon this another layer of articles is laid, and so continue until the box is filled, when it is placed in a room having a temperature from 60° to 80° C. (140° to 176° F.) and left there from 3 to 8 hours, according to the size of the box.

During this time the animal charcoal absorbs the odor, and the articles become entirely deodorized. But they must be stored in a special room, as they would again absorb the odor if stored with articles not deodorized. In the course of time, the animal charcoal loses its deodorizing ability and must be replaced by fresh material, but its deodorizing power can be restored by calcining it. This is done by placing it in sheet-iron cylinders about 50 centimetres (19.7 inches) long, and closed on both ends by tight-fitting covers. The upper lid should have a hole the thickness of a straw to allow the gases to escape. The animal charcoal, after it has been thoroughly calcined, must be allowed to cool off before the cylinders are opened, as it would be consumed if exposed to the air while hot.

Desulphurized Vulcanite.

We have stated that only *a small portion* of the sulphur combines at first *chemically* with the caoutchouc, the remainder of the sulphur used being only *mechanically* mixed with it. This sulphur becoming effective in time, causes the articles to become brittle and hard after long storage. This defect will especially show itself in vulcanized rubber hose, which becomes so hard that it will break when an attempt is made to bend it.

To prevent the vulcanite from losing its valuable properties, it is subjected to a treatment called "*Desulphurizing.*" This is done by boiling the vulcanite in caustic soda, which gradually dissolves the free sulphur, but does not attack the actual vulcanite, *i. e.*, the chemical combination of caoutchouc and sulphur.

The time required for boiling depends on the strength of the caustic soda used, and on the quantity of free sulphur which may be present. The best plan is occasionally to take a piece of vulcanite from the boiler, and test by it the progress of desulphurization. The color of desulphurized vulcanite very much resembles that of ordinary caoutchouc. Therefore, as long as the interior of the piece shows the peculiar grayish coloring of ordinary vulcanite, the mass is not properly desulphurized, and the boiling must be continued.

Vulcanite, properly desulphurized, only requires washing and drying after it has been taken out of the caustic soda, and is then a product which deserves to be called the most finished of all caoutchouc preparations, as it not only remains entirely soft and elastic in all temperatures, but does not become hard even if stored for a long time, and has no smell whatever.

There can be no doubt that among all the varieties of rubber, desulphurized vulcanite is best adapted for manufacturing articles for surgical or scientific purposes, or to be used in the nursery and sick-room (feeding tubes, air-cushions, etc.). Our experience is that it is an excellent material for hose, for conducting illuminating gas, as it combines great pliability with perfect impermeability to gas.

X.

PREPARATION OF HARD RUBBER (CORNITE AND KERATITE).

In describing the operation of burning the vulcanized caoutchouc we drew attention to the fact that the elasticity of the product perceptibly decreases if it is exposed to too high a temperature, and that its properties are changed in such a manner that finally a body remains possessing a very considerable degree of hardness and toughness whose features resemble in all respects those of horn, and for this reason has been called "*hornized*" caoutchouc. The designations cornite[1] and keratite[2] also refer to these properties.

Hard rubber is used in the manufacture of many articles, as combs, spindles, shuttles for spinning and weaving, cigar and match cases, valves, pumps for caustic fluids, surgical instruments, etc. It may be said, in brief, that hard rubber is an excellent sub-

---

[1] From "*cornu*," horn.
[2] From κεραϛ, horn. Keratin is chiefly found in horn, nails, hair, feathers, etc.—*Translator.*

PREPARATION OF HARD RUBBER.

stitute for horn, wood, leather, and even for glass and metal in many cases where these materials are used.

For making hard rubber the same materials are employed as for vulcanite, but differ in their proportions and treatment. The many admixtures which are used, serve only to increase the bulk without exerting any chemical influence upon the hard rubber, except in the case of shellac, which is added to augment its hardness.

We will treat of these components in future pages, and now give our attention to the preparation of hard rubber.

Much greater quantity of sulphur is used in making hard rubber than in vulcanite; frequently the amount of sulphur equals one-half of the weight of caoutchouc. The object of using more than the actually required amount is to cheapen the cost of manufacture. Antimony, trisulphide, or lead sulphide can be used with equal success instead of pure sulphur.

The zinc penta-sulphide which has been recommended by *Goodyear* is obtained by compounding a solution of white vitriol (zinc sulphate) with a solution of potassium penta-sulphide as long as a precipitate is formed; the precipitate is washed upon a filter and dried.

The caoutchouc, sulphur, and other ingredients are mixed in the same manner as in vulcanite, that is, between kneading rollers, and this is continued until an entirely homogeneous mass has been formed. As the mass contains a large percentage of foreign substances, it is not as soft as the unburned vulcanite, but has the consistency of a dough which can be shaped into all possible forms.

The hard rubber is generally rolled into sheets of the necessary thickness and modeled into the desired forms by pressing it into moulds. Small boxes, spectacle cases, etc., are shaped over solid cores. Sometimes larger sheets are rolled out, burned, and then worked like wood or horn, with a lathe, or saw and plane.

The burning of hard caoutchouc requires either one or two operations. If ordinary articles or simple plates are to be treated, one burning suffices, but for articles of a more complicated form it is advisable to subject them to two burnings at progressive stages of the work.

If the burning is to be done at one operation, the articles are placed in the burning apparatus, and heated for several hours—three to six—at a temperature of 150° C. (302° F.). Manufacturers differ in their statements about the degree of heat, but they are of little value as every practical man will soon find out for himself. We read, for instance, that especially excellent properties are given to hard rubber by first heating it for two hours at 110° C. (230° F.), then quickly raising the temperature to 150° C. (302° F.), and keeping it there for several hours.

By this process articles are obtained sufficiently burned, as a temperature of 150° C. (302° F.) suffices to change quite thick objects into hard rubber in the course of several hours, but we cannot understand the object or effect of heating them for two hours at 110° C. (230° F.), as sulphur, as is well known, only melts at 113° C. (235.4° F.), and it is scarcely worth while to talk about sulphur producing any effect before it is melted. We have made special experiments in regard to this matter, and have found that, even after the mass had been heated for several hours at a temperature of 110° C. (230° F.), by far the greater part of the sulphur could be extracted from it in an unaltered state by a solvent, this being a sure proof that no *chemical* action had taken place.

Hard rubber suffers considerable contraction in course of burning, and the shrinkage of the articles allows their removal from the moulds by a gentle tap. The shrinkage being uniform, owing to the even temperature, the articles do not lose their proportionate shape.

Articles of a not especially complicated shape may be burned without the use of a form. Flat articles may be subjected to the burning process without further preparation by laying them upon iron plates, but if other articles are to be burned without using forms, it is advisable to dust them with magnesia or powdered chalk, and to place them in sheet-iron boxes filled with fine sand in such a manner that the articles are completely covered by it on all sides. The sand prevents the articles from slagging during the first period of burning, an event which it would be very difficult to avoid without the use of sand.

Skilled workmen can burn and finish quite complicated articles at one operation by the above process, with a very small percentage of defective results. To prevent any possible failures it is advisable to divide the burning process into two, sometimes three, and even into four operations.

In this case the first burning, at a temperature of about 145° C. (293° F.) lasts but one hour. The articles then acquire a considerable degree of solidity, and can be taken from the forms to undergo inspection. The perfect pieces are immediately put back in the burning apparatus and finished without further manipulation, but those showing defective places, cracks, holes, etc., are repaired with a dough-like mass of caoutchouc, and heated for another hour, when they are again inspected and repaired, if necessary, and burned for another hour. This alternate inspection, repairing, and burning being continued until the articles can pass for finished goods.

Hard rubber, prepared with caoutchouc and sulphur only, has a black color, and takes a very high degree of polish. Articles manufactured from this material, although black, can also be colored any desired shade.

A distribution of the coloring substances through the entire mass has the advantage of considerably increasing its weight, but it also, to a great extent, injures its properties. To give hard rubber any desired color without altering its internal properties we make use of two methods which may be designated as " *dust-*

ing," and "*plating*," or "*enamelling*," both being well adapted for the purpose.

The dusting is done as follows: After the article has been shaped from the unburned material, it is thickly dusted with a finely powdered coloring substance by a bag containing it. The mould into which the article is to be pressed must first be uniformly dusted; after burning the article should show a uniform coloring. Any defect is repaired by repeating the dusting and then reburned

Articles of an entirely uniform color can be obtained by *plating* or *enamelling* in the following manner: The mass for the enamel is prepared in the usual manner with caoutchouc and sulphur, and the coloring agent is incorporated with it at the same time as the sulphur, by means of the rollers, the rolling being continued until a uniformly colored plate is completed.

For the purpose of enamelling hard rubber with this colored paste, we first change the uncolored mass into plates of a proper thickness, and also roll the colored paste into plates about half as thick as the first. If the hard rubber is to be enamelled on *one* side only, the two plates, the colored and uncolored, are laid one upon the other, rolled out to the required thickness, and the articles are then formed from this mass.

If the hard rubber is to be enamelled on both sides, an uncolored plate is placed between the colored ones. In this manner a different color can be shown on each side, and an enamel of any desired thickness can be applied. The thinner the enamel, the thicker must be the uncolored plate inclosed between the colored ones.

Articles called hard rubber frequently consist of a very small proportion of actual caoutchouc, and, as has been described, when speaking of vulcanite, various indifferent substances, as chalk, magnesia, zinc white, etc., are added for the purpose of increasing the weight. If these admixtures are intended to give a certain color to the hard rubber, care must be had not to use any coloring substance which may be affected by sulphur, as, in such a case, the coloring might turn out exactly opposite to what was intended. For this reason, all coloring substances containing lead, such as white lead, normal lead chromate, etc., must be avoided, as lead has a strong affinity for sulphur, and forms thereby black plumbous oxide. In this event, black, instead of white or yellowish-colored masses, would be obtained. What has been said about lead holds good in the same degree in regard to coloring substances containing copper.

The so-called lake color, prepared from organic substances and alumina, as well as zinc colors, may be used for coloring hard rubber without any further preparation, except that the lakes must be perfectly dry before they are worked into the mass. If they were used in a moist condition, the water contained in them would escape during the burning, in bubbles, and cause the mass to swell up, and the surface to become uneven, while the interior would be porous instead of compact.

The scraps left after forming articles from the unburned mass are immediately kneaded together, passed through between the rollers, and may then be used for preparing other articles. But the waste of hard rubber, that is to say, after it has been burned, can only be used for one purpose, namely, for manufacturing lacquer; but of this we will speak later on.

The best plan is to form the articles by pressing or stamping them from the mass, as there is much less waste than when preparing plates from which the articles are to be manufactured.

The hardness and elasticity of hard rubber principally depends on the quantity of sulphur which has been added to the caoutchouc. In the following we give a few receipts adapted for the purpose.

Articles sufficiently elastic and pliable, so that they will not break, even when sharply bent, can be made from the following composition:—

	Parts.
Caoutchouc	86 to 88.
Sulphur	14 to 12.

This is especially well adapted for manufacturing combs and such thin articles as are to possess a high degree of elasticity and considerable solidity.

By mixing together:—

	Parts.
Caoutchouc	76 to 80
Sulphur	24 to 20

a mass is obtained which, in regard to elasticity, is

nearly equal to the foregoing, but is somewhat more fragile.

The largest part of the articles, as combs, etc., sold as hard rubber are made of a composition resembling the latter, but containing more sulphur and consequently are much cheaper.

When great hardness and solidity, with but little elasticity are required, as in a material suitable for knife handles, rollers, buttons, tool handles, door knobs, lock plates, etc., the percentage of sulphur is increased and—

	Parts.
Caoutchouc	65 to 76
Sulphur	35 to 24

form the best compositions for such purposes.

It is a remarkable fact that some resins, although possessing a considerable degree of brittleness, impart a certain elasticity to hard rubber. Shellac is especially effective in this respect. It can be used either bleached or unbleached (the so-called ruby shellac). The latter is to be preferred for articles of a dark color, as it is far cheaper than the bleached article, and answers the purpose equally well.

The shellac should be powdered as fine as possible, and intimately mixed with the rubber by continuous rolling. A piece of shellac large enough to be visible to the naked eye would be sufficient to spoil the appearance of the article. Hard rubber will bear the admixture of a large quantity of shellac, and in some cases an amount equal to the caoutchouc may be used.

A composition consisting of:—

	Parts by weight.
Caoutchouc	88
Sulphur	12
Shellac	50

was formed into sticks having a cross section of 1 square centimetre (0.155 square inch) which could be bent to a considerable extent after they had been burned, but were at the same time so elastic that the sticks always sprung back to their original straight form. The rubber was so hard that thin shavings could be cut from it only with a very sharp knife. In consequence of these properties this variety of hard rubber furnishes an excellent material for the manufacture of bobbins, and shuttles, as these can be made so thin that their sides are scarcely thicker than thin pasteboard.

Hard rubber shows as much indifference to chemical influences as vulcanite does, but this condition of the material exists only when it is compounded with sulphur alone. Compositions containing chalk, magnesia, etc., in addition are affected by acids and alkalies.

Hard rubber can be used with great advantage for making bowls for silver baths, spatulas, rollers and other utensils used by photographers, also for stoppers and caps of bottles containing corrosive fluids.

XI.

PREPARATION OF ARTIFICIAL IVORY, EBONITE, EBURITE, OR "IVOIRE ARTIFICIEL."

For many years chemists have endeavored to prepare a substance to serve as a substitute for ivory, which is becoming scarcer and dearer every year. Their attention was principally directed to compositions with glue as a base, to which were added finely powdered white substances, and such bodies as would make the glue insoluble; the salts of alumina and tannin being chiefly used for this purpose. It cannot be denied that the process of preparing such masses has been successful as far as their external appearance is concerned, it being scarcely possible to distinguish them from genuine ivory. But they lack one of the principal properties of the genuine article, namely that of combining great elasticity with solidity.

It has been tried to use caoutchouc for preparing a mass which would possess as nearly as possible all the properties of ivory, and these experiments have been so successful that substances are now obtained which can be employed for manufacturing a large number of articles formerly made of ivory. But they cannot take the place of the genuine material where elasticity combined with great solidity is a requisite. Many attempts have been made to manufacture billiard balls from such composition, but they have never been very successful. The balls in a short time become full of cracks, and frequently break even if but gently struck.

To prepare an elastic ebonite, there must be added to the composition a certain percentage of pure caoutchouc, but this prevents the ebonite from having an entirely light color.

We desire to mention that a large number of preparations are sold under the name of ebonite, which contain but little of caoutchouc or gutta percha, foreign substances, giving weight, being the chief components. The same may be said of compositions sold as vulcanite, which sometimes contain less than 33 per cent. of caoutchouc.

All imaginable bleaching agents have been tried and recommended to completely decolorize and bleach caoutchouc, but no process has been entirely successful, and it can justly be said that a complete bleaching of it is not possible with the means at our command. If caoutchouc is treated with a bleaching agent exerting any effect upon it, it will acquire a lighter color (it becomes a very light yellowish brown), but a *chemical* change will at the same time take place, and the mass bleached by chlorine cannot be called caoutchouc any longer.

Several methods have been made public, by which it is claimed that caoutchouc can be bleached without undergoing a chemical change. We have subjected all these methods to a test and always obtained the same result. *The bleached product did not possess the properties of caoutchouc.* Nearly all these methods amount to the same thing, namely, that the caoutchouc is allowed to swell up very much (a complete solution is not necessary), and chlorine gas be introduced into the swelled up mass. Chloroform, bisulphide of carbon, benzol or oil of turpentine are recommended as solvents, but we found bisulphide of carbon, benzol and *rectified petroleum* the most suitable solvents.

A special apparatus is required for treating caoutchouc with chlorine. This consists of a wooden vat lined with lead. It should have a lid lined with lead, and capable of being screwed down air-tight. In the centre of the lid or cover is a revolving shaft connected with a stirring apparatus made of leaden rods. The pipes conducting the chlorine reaches to the bottom of the vat, and a funnel provided with a stop-cock is fitted into the cover to admit new fluid when required. A small pipe is fitted into the cover and passes into a tank of water adjoining the vat, for the escape of the chlorine gas. The accompanying illustration (Fig. 3) represents the entire arrangement of the apparatus. The chlorine enters through C ; R is the stirring apparatus ; J the funnel ; A the escape pipe for the chlorine ; H is a stop-cock in the bottom of the vat.

The caoutchouc, after being cleansed and cut in small pieces, or, still better, in shreds, is placed in the vat which is then closed, and the solvent poured in through the funnel J. The stirring apparatus R is started and kept in motion until the solution is complete. Then the chlorine is introduced and allowed to flow until its escape is noticed from the pipe A.

Spirit of wine equal in volume to that of the solvent is introduced into the vat through the funnel J. By this the bleached caoutchouc is separated and precipitated in a slimy condition. The solution is constantly

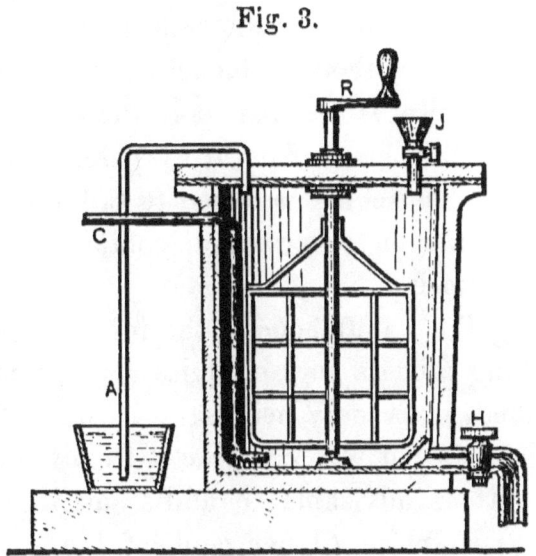

Fig. 3.

stirred to permeate it with the spirit of wine. The stop-cock H is then slightly opened to allow the fluid to run off. The spirit of wine and original solvent are finally separated by distillation.

As has been stated, caoutchouc bleached in this manner is of a brownish-yellow color, and should be immediately worked, as it is quite soft when it comes from the bleaching apparatus, and can be easily compounded with admixtures.

By an American bleaching process, the caoutchouc is dissolved in chloroform, and the solution allowed to clear by settling, then treated with ammoniac gas until all the caoutchouc, appearing in the form of a sponge, is separated. It is then washed in hot water to free it from all traces of chloroform, and subjected to further treatment. This method furnishes no better results than the former, and is rather expensive.

The simplest method of bleaching caoutchouc for preparing ebonite is to treat it in the usual manner, and to form thin bands from it by passing it through the rollers. These are allowed to fall into a roomy vat provided with a cover which contains water saturated with chlorine.

By using the caoutchouc in the form of thin bands the bleaching process goes on quite rapidly, and when accomplished, it is only necessary to wash the mass several times in hot water to free it from the adhering chlorine. It is advisable to add a small quantity of sodium hyposulphite (1 per cent. of the salt is sufficient) to the first washing water, as this salt has the effect of removing every trace of chlorine which may still be present. If any of the chlorine were to remain it would exert an injurious effect afterwards when the material has to be worked in machines constructed of metal.

The best plan is to subject the bleached mass to further treatment immediately after it has been washed, as it is then very plastic. But if it is not desired, nor possible to use the mass at once, it is advisable to moisten it with some bisulphide of carbon or benzol and to let it remain for some time in a hermetically closed vessel. A small quantity of the solvent will cause the mass to swell up somewhat, and it can then be worked with greater ease.

The next labor required for preparing ebonite is to incorporate several substances with the plastic mass. Either whiting, oxide of zinc or white lead is used for producing a white material. Artificially prepared sulphate of baryta (the *blanc* or *fix* permanent white used as a coloring substance by the manufacturers of wall papers) is also a very suitable material for the purpose. Articles colored by pigments containing white lead, lose their white color in time and turn gray.

The colored masses are worked in different ways either direct or indirect. If the direct method is chosen, the ebonite is immediately pressed into hot moulds, but sharp impressions can only be obtained by using a very high pressure. Many elegant articles of ebonite can be prepared in an inexpensive manner by this method, provided it is not absolutely necessary that the substance should be entirely homogeneous. Knife handles, buttons, etc., may be prepared by direct pressing.

For solid, but at the same time non-elastic articles, it is best to use the indirect method, that is to say, the ebonite is formed into cubes, and the articles are shaped from them by the lathe, etc. Therefore the preparation of ebonite must be regulated by the use intended. For instance ebonite for billiard balls should undergo less pressure than that for making thin plates.

The greater part of the numerous receipts given

for preparing ebonite, are of no special value, as its quality depends so much upon the caoutchouc used, the duration of the chlorine treatment, etc. According to many of the receipts masses are made resembling vulcanite, by using sulphur and gutta-percha in addition to caoutchouc, and heating them to 150° or 160° C. (302° to 320° F.).

Jacobsen gives an American receipt, according to which an ebonite mass consists of the following substances:—

	Parts.
Caoutchouc	100
Sulphur	45
Gutta-percha	10

The mass is heated to 157° C. (314.6° F.).

XII.

CAOUTCHOUC COMPOSITIONS.

CAOUTCHOUC belongs to that class of substances which admit of combinations with the most diverse materials, and, consequently, compositions can be made that are serviceable for many purposes. Some of the most prominent are *Kamptulicon* (*caoutchouc leather*), *balenite* (*artificial whalebone*), *plastite* (*a hard non-elastic, but easily formed mass*), composi-

tions for sharpening knives, caoutchouc enamel and *lacquers* for coating metals, wood, etc.

Many of these compositions are considered trade secrets, and the respective manufacturers do every thing in their power to keep up this idea; but there is actually no secret about them, as it is of no consequence whether one or the other substance is added to the mass, as long as the principal properties of the composition remain. In the following we will briefly describe the most important of these compositions, and leave it to the intelligence of the manufacturers to change the properties of the masses by a corresponding change in their composition:—

KAMPTULICON.

The composition known by this name is of English origin, and is especially well adapted for the manufacture of floor-cloths subjected to hard usage, for coating articles, etc. Genuine kamptulicon consists of an intimate mixture of caoutchouc and powdered cork, and is prepared in the following manner:—

Waste of cork, and old corks also, is taken and cleansed by washing it several times in water. The washed and well-dried mass is comminuted by grating it upon a drum provided with small teeth like a rasp, and then ground into a fine powder.

The caoutchouc is cleansed in the usual manner, and rolled into thin bands between closely-set rollers. The bands are strewed uniformly with the powdered cork, and then subjected to further treatment. This is done in the same manner as described for preparing the vulcanite mass; that is, by rolling, kneading, and repeated rolling until an entirely homogeneous mixture has been formed. Finally, plates two to five millimetres (0.0787 to 0.1968 inch) thick are formed, and these are covered either on one or both sides with a coat of good linseed-oil varnish or oil paint. Of course, with the assistance of oil paint, plates with various patterns (carpet and parquet designs) can be produced.

Powdered sulphur may be also incorporated with the caoutchouc besides powdered cork, and the articles may be subjected to the burning process after they have been formed, and in this manner a vulcanized kamptulicon is obtained.

The principal advantages of kamptulicon are, that with very little weight it combines great elasticity, and, for this reason, is well adapted for floor covering in passage ways, where the noise made by walking is to be prevented as much as possible.

Kamptulicon can be used as a block cushion under stamping presses to weaken the shock, but it should be inclosed in an iron ring to prevent it splitting. It serves also to make wheels for polishing brass, steel, German silver, and other metals. This is done by covering a wooden disk with a piece of kamptulicon of the proper size.

A uniform color can be made to permeate the entire mass, which can be worked into a kind of mosaic in floor coverings. This condition is obtained by incorporating the coloring matter with the caoutchouc. Colcothar, ultra marine, lamp-black, etc. are used for this purpose. The masses colored by any of the above substances are rolled out, and it is then a very easy matter to cut stars or other designs from them by means of a sharp knife or a suitable stamp, and to combine these into any desired pattern. As the matter is colored through and through, floor-cloths, or other articles manufactured in this manner, retain their beauty as long as the articles themselves last.

CAOUTCHOUC LEATHER.

In most cases this is identical with kamptulicon, although sometimes it is manufactured in a different manner. While genuine kamptulicon is always composed of caoutchouc and cork, caoutchouc leather frequently contains, instead of cork, any kind of fibrous substance, such as hemp, flax, jute, etc. It is generally manufactured in the following manner: Caoutchouc (mostly in the form of small waste pieces, of which there will always be large quantities in rubber factories) is either entirely dissolved by a solvent or at least allowed to swell up very much; the fibrous substance is then incorporated with the mass, and the latter is made homogeneous by long-continued rolling.

The easiest way of incorporating the fibrous substance is by stirring as much of it into the half-fluid mass of caoutchouc and solvent (purified petroleum is the best for this purpose) as can conveniently be done. The mass is then placed upon a table, which is quite thickly strewn with fibrous substance, and rolled into a cylinder. When in this manner a mass has finally been obtained, which possesses sufficient consistency to allow of it being worked between the rollers, the incorporation of fibrous substance is continued there until a sufficient quantity of it has been kneaded in, to impart a suitable degree of solidity to the mass.

It is advisable to repeatedly form the bands, which have been obtained by rolling, into lumps, and to pass these again through the rollers, as by these means the fibres are piled in different directions (forming, so to say, a kind of felt), and by doing this the solidity of the substance will be considerably increased.

As far as solidity and tenacity are concerned, caoutchouc leather surpasses by far the kamptulicon; but the latter is softer and more elastic. Thus far, kamptulicon, as well as caoutchouc leather, is too expensive; but when these valuable substances can be brought into the market at cheaper prices, there can be no doubt that the demand for them, and their use for various purposes will greatly increase.

Balenite,

or artificial whalebone, is a substance, as the name indicates, intended to serve as a substitute for genuine whalebone. A mass which shall answer the intended purpose must possess considerable elasticity as well as solidity—must, therefore, be a medium between vulcanite and hard rubber. A mass answering the purpose very well is prepared according to the following receipt:—

	Parts by weight.
Caoutchouc	100
Ruby shellac	20
Burned magnesia	20
Sulphur	25
Orpiment	20

The foreign substances are incorporated with the caoutchouc, the mass is pressed into forms, usually balenite is formed into plates or prismatic bars, and burned at a moderate heat. The mass which is obtained may serve in all cases as a substitute for the genuine whalebone, and may also be used for bobbins, etc. On account of its light weight and indestructibility, balenite may be highly recommended for the manufacture of gunstocks, as also of elastic plates and splints for surgical purposes.

Plastite.

The so-called plastite is a mass resembling hard rubber, but differing from the latter in possessing no elasticity, although a considerable degree of hardness. As plastite can be brought into any desired form, and is composed of a large percentage of substances of little value, it is especially well adapted for the manufacture of pressed ornaments, small frames, boxes, heels of shoes, etc.; in short, for all purposes for which wood, metal, horn, etc., are used.

The so-called coal-tar asphaltum is an important component of plastite; this substance, which forms a deep black, shiny and hard mass, is gained in the distillation of coal-tar, as residuum after all volatile substances have been distilled off. Besides coal-tar asphaltum, sulphur, and magnesia, and sometimes orpiment form a part of plastite.

Magnesia can be very well replaced by other indifferent substances, such as finely powdered and washed chalk, etc., but the use of magnesia offers the advantage that the masses can be made of great volume, and at the same time of little weight, as magnesia is very light.

A plastite mass possessing very good properties may be prepared according to the following receipt:—

	Parts by weight.
Caoutchouc	100
Sulphur	20 to 25
Magnesia	40 to 50
Orpiment	40 to 50
Coal-tar asphaltum	50 to 60

The foreign substances are incorporated with the caoutchouc in the usual manner; the separate articles are pressed in hot iron moulds and are then burned. On account of its great hardness and solidity, plastite takes a high degree of smoothness and polish, and for this reason is well adapted for the manufacture of handles for umbrellas and canes, door-knobs, etc.

GRINDING AND POLISHING COMPOSITIONS.

Caoutchouc possesses the specific property of holding foreign substances, once incorporated with it, very tenaciously. If the incorporated substances are hard, the mass is suitable for grinding or sharpening; if soft, the composition serves for polishing. The first category includes powdered pumice-stone, powdered glass, quartz, sand, emery. To the second class belong colcothar, graphite, talc.

There are several receipts for preparing grinding compositions, which have been highly recommended, especially for sharpening and polishing knives. We give them in the following :—

I.

	Parts by weight.
Caoutchouc	280
Powdered emery	1120
Lampblack	$6\frac{1}{3}$

II.

	Parts by weight.
Caoutchouc	280
Graphite	512
Lampblack	$6\frac{1}{3}$

III.

	Parts by weight.
Caoutchouc	280
Graphite	488
Lampblack	$6\frac{1}{3}$

IV.

	Parts by weight.
Caoutchouc	280
Zinc-white	84
Yellow ochre	1120

V.

	Parts by weight.
Caoutchouc	280
Sulphur	84
Powdered emery	1120

As will be seen from the receipts, I. and V. contain emery, which, on account of its hardness, serves as a grinding agent. The addition of lampblack in the other compositions is not essential, as the only object of it is to give a black color to them. Nos. II. and III., on account of the percentage of graphite they contain, must be considered as polishing compositions; and No. IV. also possesses the same property.

We have endeavored to prepare compositions which shall answer for one or the other purpose; but one suitable for grinding, and at the same time for polishing, can only be prepared by using a hard body in the form of an impalpable powder, equalling in fineness the very finest flour.

Graphite or talc, which, of course, must be powdered as finely as possible and washed, are especially well adapted for polishing compositions. The caoutchouc is mixed with 150 to 200 per cent. of this powder, and the entire mass is vulcanized by adding 10 to 15 per cent. of the weight of caoutchouc of sulphur to it, and subjecting it to the burning process.

Grinding compositions may be prepared by using powdered glass, pumice-stone, flint, or emery; the masses containing powdered glass or pumice-stone (being the softest) may be used for grinding brass or bronze, while those containing powdered flint for grinding steel; the masses containing emery may be even used for grinding precious stones, as emery is the hardest body next to the diamond.

To change the hard bodies—glass, flint, and emery—into fine powder, it is necessary to make them red-hot, and to throw them while in this condition into cold water. They become very brittle in consequence of the quick cooling off, and can then be ground into fine powder without great difficulty.

If the grinding composition is to be subjected to considerable wear, it is advisable to add to the caoutchouc, besides the powder of the hard body, some sulphur, and to burn the mass sufficiently to change the caoutchouc into hard rubber.

It is merely a matter of choice what form is to be given to the grinding and polishing compositions.

Revolving circular disks, against which a piece of vulcanized caoutchouc is pressed, are very suitable for sharpening and polishing table knives. If a knife is placed between the disk and the piece of caoutchouc it will appear polished or ground after a few revolutions of the disk.

For manufacturing purposes, especially for metal workers, it is best to give to the grinding or polishing masses the form of ordinary grindstones, that is, that of circular disks.

The quantity of powdered hard substances to be incorporated with the caoutchouc, may be a very large one, especially if hard rubber is used and may amount to as much as four times its weight.

Caoutchouc Enamel.

Hard rubber, on account of its elasticity and solidity, is well adapted for coating articles of metal which are to be protected against rust. For the purpose of coating metal with a thin layer of hard rubber, it is brushed over with a solution of caoutchouc in benzol or petroleum, and is then dusted with powdered sulphur. Both operations are repeated after the first coat has become dry. The articles coated in this manner are quickly heated to a temperature of from 160° to 170° C. (320° to 338° F.), when the well-known reciprocal action between sulphur and caoutchouc takes place, and they will come out with a coat of hard rubber. Defective places in the coating can be repaired by repeating the brushing over with the caoutchouc solution, dusting with sulphur and burning.

If it is desired that the coat should show an entirely uniform black color, it is advisable to dust the article with a fine black pigment after it has been dusted with sulphur. The so-called vine-black can be especially recommended for this purpose, as it forms an entirely dry powder which can be easily and completely dusted away, which is not the case with the majority of black pigments, for instance, lamp-black, as more or less tarry matter always adheres to them.

The following process may be recommended for preparing colored enamels of somewhat greater thickness. An entirely clear, but rather thick solution of caoutchouc is prepared. This is intimately mixed with about 12 per cent. of the weight of the originally used caoutchouc of the finest powdered sulphur and the coloring substance to be used. The mass obtained in this manner should have a consistency equal to thick oil paint; should it be too thick to allow of it being evenly applied with a brush, it may be reduced with oil of turpentine, or, in case it is too thin, this may be remedied by an addition of coloring matter.

If benzol or bisulphide of carbon is used as a solvent, it will be very difficult, on account of the great volatility of these fluids, to evenly apply the mass with a brush; it is therefore best to allow the caoutchouc to swell up strongly in benzol or bisulphide of carbon, and effect the complete solution with oil of turpentine or rectified petroleum.

Bristle brushes are used to coat the articles with the composition, and it should be done in thin but frequently repeated applications. If a white ground mass is used marbled designs can be produced by using yellow, red, or blue. The beauty of the work will of course depend on the skill of the workman.

When the entire coat is finished it is dried, which can be accelerated by exposing the article to a temperature not exceeding 100° C. (212° F.). In case the enamel shows defective places it is repaired and finally burnt in at a temperature of 160° C. (320° F.). The caoutchouc enamel manufactured in this manner adheres very tightly to metal and will take a very high degree of polish. As it will stand a temperature of over 200° C. (392° F.), it can be advantageously used for enamelling the exterior cases of the so-called *Meidinger's* furnace.

XIII.

CAOUTCHOUC LACQUERS.

The properties of caoutchouc, especially its elasticity and chemical indifference, make it particularly well adapted for the manufacture of varnishes and lacquers, the latter being more extensively used every year by the trades requiring these materials. Certain waste which could not be used in any other manner can be very advantageously employed in rubber factories for the manufacture of lacquers.

Caoutchouc must be first transformed into a solution before it can be manufactured into varnish or lacquer. The solvents which have so frequently been mentioned in this work are very well adapted for this purpose. But in preparing such solutions it becomes necessary to observe many things, so as to obtain them of uniform quality, and as they are used for many other purposes, we will first discuss the manner of preparing them.

Preparation of Caoutchouc Solutions.

Solutions of caoutchouc for manufacturing purposes are always prepared on a large scale, and in iron vessels hermetically closed. The caoutchouc is cut up in small pieces, and the solvent should be as free from water as possible.

Although the different varieties of caoutchouc generally show the same composition, they differ very much, as experience has shown, in regard to their behavior towards solvents. While, for instance, one variety can be easily dissolved in oil of turpentine, it is very difficult to dissolve another with the same solvent. It is therefore advisable to execute a preliminary test with small quantities of the caoutchouc and solvent.

The most favorable results are always obtained by adding from 5 to 40 per cent. of highly rectified spirit of wine to the solvent. It is very likely that the good effect produced by this addition is due to the property of spirit of wine of absorbing water.

Masses of any consistency, from solid caoutchouc to a complete solution, may be prepared in this manner as the caoutchouc, as has been mentioned, swells up very strongly in the solvent before it dissolves.

If bisulphide of carbon or benzol is used for preparing caoutchouc solutions, it is necessary to use special precaution to prevent volatilization. The vessel should have a broad rim upon which a strip of vulcanite is laid, and the rim of the cover fastened down tightly upon it. The vessel is also provided with a stirring apparatus, and in its entire arrangement very much resembles that used for bleaching caoutchouc, (see Fig. 3.)

The dissolving powers of most fluids can be increased by heat, and the dissolution of caoutchouc can be much accelerated by placing the apparatus in a boiler filled with water, which is heated. If bisulphide of carbon is used the temperature should not exceed 40° C. (104° F.), if benzol to 60° C. (140° F.), and for oil of turpentine or rectified petroleum to 100° C. (212° F.), but this heating should only be done towards the end of the operation.

Generally from 24 to 30 hours are required for the accomplishment of a complete solution, but as the time can be considerably shortened by heating and repeatedly stirring the mass, it is advisable to arrange the stirring apparatus in such a manner that it is connected by a pulley with the engine of the factory and can be kept in a constant slow rotation.

When the fluid which is formed after the solvent has acted upon the caoutchouc for a sufficiently long time, is examined in a glass vessel, it will be observed that it is never uniform, but that small lumps of more or less swollen caoutchouc float in the thick solution. These can never be completely dissolved even by using very large quantities of the solvent.

Therefore, to obtain uniform solutions, the mass coming from the apparatus must be subjected to mechanical treatment, and this consists in kneading or squeezing it between rollers. If bisulphide of carbon or another volatile solvent has been used for dissolving the caoutchouc, this must be done in a hermetically closed vessel to prevent heavy losses of solvent by evaporation.

Fig. 4 represents an apparatus which is particularly well adapted for kneading caoutchouc solutions; it is arranged in the following manner:—

Two smooth rollers, a and b, of equal diameter, lie in a box K of wood or iron. The rollers are acted upon by cog-wheels on the outside of the box in such a manner that they revolve at unequal velocities, and are placed so close together as to leave only a very narrow space between them. The solution of caoutchouc to be kneaded is poured in the sheet-iron vessel G, in the bottom of which is a slit running parallel with the rollers. Smoothing blades S, pressing closely against the rollers, are placed on both sides of them.

The solution of caoutchouc which is scraped by the blades from the rollers flows into a collecting vessel, and eventually upon a second, third, or fourth pair of rollers.

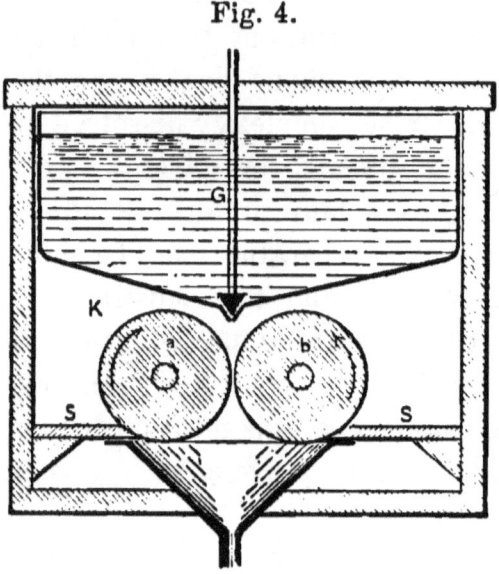

Fig. 4.

To control the flow of solution from the vessel G upon the rollers, there is a metallic wedge fitting the slit in the bottom of the vessel G. This wedge can be raised and lowered by a rod projecting above the box, and the discharge of the solution regulated thereby. So as to be able to observe the flow of the caoutchouc solution, a pane of glass is fitted into the side of the box.

By repeatedly treating the partly dissolved and partly swollen mass of caoutchouc between the rollers, an entirely homogeneous mass is finally obtained, the consistency of which will of course depend on the proportions of caoutchouc and solvent. Very thickly-fluid solutions can be used for casting different articles in hollow moulds, somewhat thinner solutions for pasting together pieces of caoutchouc, etc.

By reducing the swollen masses of caoutchouc, which have been made homogeneous by passing them through the rollers, with ether, chloroform, oil of turpentine, etc., fluids are obtained which can be used, as varnishes without further preparation. In drying, they leave behind a very thin, nearly colorless film of caoutchouc, such solutions being well adapted for coating copper plates and maps, which can then be cleansed with a moist sponge.

For many purposes, caoutchouc solutions are not used by themselves, but compounded with copal varnish, boiled linseed oil, dammar resin, etc. Varnishes prepared with an addition of resin-varnishes, or lacquers, show a strong lustre, while pure caoutchouc varnishes (that is solutions of pure caoutchouc) possess scarcely any lustre whatever.

In the following we give a few receipts for preparing caoutchouc varnishes for different purposes.

Caoutchouc Varnish for Leather.

Caoutchouc	1	kilogramme	(2.2 lbs.)
Dissolved in turpentine	8	"	(17.6 lbs.)
Mixed with fat copal lacquer	6	"	(13.2 lbs.)
Boiled linseed oil	4	"	(8.8 lbs.)

Caoutchouc Varnish for Gilders.

Caoutchouc	1 kilogramme		(2.2 lbs.)
Dissolved in rectified petroleum	8	"	(17.6 lbs.)
Mixed with copal lacquer	4	"	(8.8 lbs.)

Caoutchouc Varnish for Glass.

	Parts by weight.
Caoutchouc	1
Dissolved in chloroform	60
Mastic	10

This varnish, which adheres excellently on glass, can be colored as may be desired, and with it imitations of flashed glass can be prepared, and glass cemented to glass. It is also well adapted for fastening letters of glass or metal upon glass.

MARINE GLUE.

The so-called marine glue is an excellent caoutchouc lacquer for protecting wood and metal against the action of water. It consists of a solution of 1 part of caoutchouc in 12 parts of rectified petroleum which are combined, by heating and stirring, with 6 parts of shellac or asphaltum. It should be applied at a temperature of from 130° to 140° C. (266° to 284° F.).

Marine Glue for Damp Walls.

It is a well-known fact that it is a very difficult matter to keep the basement walls of houses entirely dry. Generally they absorb so much moisture from the ground that the glue which has been mixed with the paint commences to mould, and the painter's work falls off, or in case the room is papered the wall-paper puffs up and becomes stained. These evils can be best avoided by using the following marine glue:—

	Parts.
Caoutchouc	10
Whiting	10
Oil of turpentine	20
Bisulphide of carbon	10
Colophony	5
Asphaltum	5

These substances are put in a large bottle, and this is closed as air-tight as possible. It is then put in a moderately warm place and allowed to stand until the soluble substances have become dissolved; this can be hastened by frequently shaking the bottle. The wall to be dried is first thoroughly cleansed, the glue is then applied with a flat brush, and should be laid on about 20 to 30 centimetres (7.86 to 11.79 inches) higher up than the wall appears to be damp. Paper which will adhere very tightly to it is then laid over the glue while it is still sticky.

The wall-paper can be immediately pasted upon this paper, and if the glue has been prepared with due care will never fall off, as the wall will always be dry.

Jeffery's Marine Glue.

This consists of—

	Parts.
Caoutchouc	3
Undistilled coal tar	36
Asphaltum	6

The caoutchouc is cut up in small pieces and dissolved in the undistilled coal tar, and the asphaltum is then added.

The marine glue obtained in this manner is so hard that it cannot be easily melted over an open fire. When it is to be used it is first softened in a water bath and can then be made more fluid over a coal fire without running the risk of scorching it.

Hard Rubber Lacquer.

Waste and broken articles occurring in the manufacture of hard rubber can be used for one purpose only, namely, for the fabrication of a lacquer.

The pieces of hard rubber are melted in an iron pot and must be constantly stirred to prevent it from burning. The melted mass is poured in a thin stream upon iron plates, where it congeals to a brittle mass resembling asphaltum. The latter is broken into pieces, and put in a bottle, and rectified petroleum, or, what is still better, benzol is poured over them. The quantity of solvent added must be sufficient to produce a fluid which can be easily applied with a brush. The fluid is allowed to stand for a considerable time, during which the foreign substances mixed with the hard rubber, and which are insoluble in petroleum or benzol, settle to the bottom, and the solution is then poured off very carefully.

Hard rubber lacquer when applied to wood or metal forms a brownish-yellow to black coating which strongly resists atmospheric influences, and for this reason is especially adapted for varnishing machines erected in the open air.

Caoutchouc Cements.

Caoutchouc Cement for Glass, No. 1.

	Parts.
Caoutchouc	1
Mastic	12
Dammar	4
Chloroform	50
Benzine	10

Caoutchouc Cement for Glass, No. 2.

	Parts.
Caoutchouc	12
Chloroform	500
Mastic	120

This cement, when applied to glass, adheres at once, and, when dry, possesses a high degree of elasticity.

Caoutchouc Cement for Glass, No. 3.

	Parts.
Caoutchouc	2
Mastic	6
Chloroform	100

The solution takes place by allowing the mass to stand for several days in a cold place. This entirely transparent cement must be applied quickly, as it becomes thickly fluid in an uncommonly short time.

Caoutchouc Cement for Rubber-Shoes.

Good rubber-shoes consist of a firm tissue which has been made waterproof by a coating of caoutchouc. The following cement is used for repairing holes in them:—

	Parts.
A.—Caoutchouc	10
Chloroform	280
B.—Caoutchouc	10
Colophony	4
Turpentine	2
Oil of turpentine	40

The solution A is prepared by allowing the caoutchouc and solvent to stand in a bottle. The solution B is prepared by melting the finely-cut caoutchouc with the colophony; the turpentine is then added; and, finally, the entire mass is dissolved in the oil of turpentine. The two solutions are then poured together.

If a hole in a rubber-shoe or waterproof coat is to be repaired, a piece of close linen is first dipped into the cement and laid upon the place to be repaired, which also has been brushed over with the cement. As soon as the linen adheres the cement is applied to the torn place and smoothed. By using some skill the place may be repaired in such a manner that not a trace of the hole can be detected.

Soft Cement of Caoutchouc and Lime.

	Parts.
Caoutchouc	150
Tallow	10
Slaked lime	10

The caoutchouc is cut up into small pieces and gradually incorporated with the tallow, which should be melted in a small brass pan. When all the caoutchouc has been added, the mass is heated, under constant stirring, until the caoutchouc is melted. A well-fitting cover should be held in readiness to be able to extinguish the flame at once in case the caoutchouc should ignite. The lime is stirred in only after the caoutchouc has been entirely melted.

This cement is particularly well adapted for sealing bottles in which corrosive substances, such as nitric acid, etc., are to be transported. The lime may be entirely omitted if it should be desired to make the cement still more indifferent to chemical influences. Caoutchouc and lime cement remains always tenacious, and is, therefore, well adapted for cementing substances which are exposed to repeated shocks.

Hard Caoutchouc Cement.

	Parts.
Caoutchouc	150
Tallow	10
Minium	10

This cement is prepared in the same manner as the one described above. The minium imparts a red color to it, and has the effect that, after some time, it congeals to a hard mass.

For a better view we will add here receipts for preparing cements with gutta percha.

GUTTA PERCHA CEMENTS.

Gutta percha Cement for Glass.

	Parts.
Gutta percha	100
Black pitch or asphaltum	100
Oil of turpentine	15

This cement, which must always be used while hot, is well adapted for all substances to be cemented, and adheres especially well to leather.

Gutta percha Cement for Leather.

A solution of gutta percha in bisulphide of carbon of the consistency of syrup, and sufficiently reduced with petroleum, does excellent service for this purpose.

The cement is quickly applied in a thin layer, and the pieces of leather are pressed tightly together.

Cement for Rubber Combs.

A. Bleached gutta percha is changed into a very thick solution by bisulphide of carbon.

B. Sulphur is dissolved in bisulphide of carbon.

The parts to be cemented are brushed over with solution A, and pressed tightly together. When dry, the cemented place is brushed over with solution B.

Elastic Gutta percha Cement.

	Parts.
Gutta percha	10
Benzine	100
Linseed oil varnish	100

The gutta percha is dissolved by itself in the benzine and the solution, after it has become clear, is poured into a bottle containing linseed oil varnish, and both fluids are mixed together by shaking. This cement is very elastic and may be used for making tissues waterproof. For this purpose it is applied with a broad brush, and when this is done uniformly it forms an entirely colorless coating of a beautiful lustre.

Leather may be also cemented together with this cement, and it is especially well adapted for cementing soles on shoes, as it is so elastic that it does not break if bent ever so much. For the purpose of making it adhere to the leather as tightly as possible, the side of the leather to which the cement is to be applied should be well roughened.

Asphaltum Cement for Leather Straps.

	Parts.
Asphaltum	12
Colophony	10
Gutta percha	40
Bisulphide of carbon	150
Petroleum	60

The asphaltum, colophony, and gutta percha, are placed in a bottle which stands in boiling water and treated for several hours with the petroleum. The solution is allowed to cool off, the bisulphide of carbon is added and the entire mass is allowed to stand for several days, but should be frequently shaken in the meanwhile. The cement is applied in a uniform layer to the straps. They are then passed between warm rollers and subjected to a strong pressure when they will adhere together very tenaciously.

Gutta percha Cement for Horses' Hoofs.

For filling up fissures and cracks in the hoofs of horses a cement is required which possesses elasticity and solidity, and at the same time the property of resisting the action of water. A mass answering this purpose in all respects consists of:—

	Parts by weight.
Gum ammoniac	10
Gutta percha (purified)	20 to 25

The gutta percha is heated to 90° or 100° C. (194° or 212° F.), and worked together with the finely powdered gum until a homogeneous mass has been formed. When the cement is to be used it is first heated until it becomes soft, and is introduced with a knife into the fissure or crack of the hoof, but the place to be cemented must be first thoroughly cleansed. As soon as the cement cools off to an ordinary temperature it becomes solid, and so hard that nails can be driven into it.

Caoutchouc and gutta percha form an essential part of a large number of varnishes and lacquers as well as of cements. Their composition and the manner of preparing them have been very accurately and thoroughly described in the " *Treatise on the Fabrication of Volatile and Fat Varnishes, etc.*" From the German of Erwin Andres, with additions. Translated and edited by Wm. T. Brannt, Philadelphia, Henry Carey Baird & Co., 1882; and we therefore refer our readers who may desire further information about this subject to the above-mentioned work.

HEVEENOID.

A caoutchouc mass comes now into commerce under the name of " *Heveenoid,*" which is claimed to be more pliable, durable, and insoluble than any other composition. It is composed of caoutchouc, camphor, and sulphur. *Soft* heveenoid consists of:—

	Parts.
Caoutchouc	2
Camphor	2
Lime	$\frac{1}{16}$
Sulphur	$\frac{1}{2}$

Hard hevcenoid consists of:—

	Parts.
Caoutchouc	3
Camphor	2
Glycerine	$\frac{1}{2}$
Sulphur	8

It has been invented by *Henry Gerner*, of New York, and has been patented in this country as well as in Europe. The *Heveenoid Manufacturing Co.*, of New York, works under the American patent.

Metallized Caoutchouc.

According to the "*Moniteur Industriel*," 1880, vol. 7, p. 64, this is prepared by the *Franco-American Rubber Co.*, by mixing caoutchouc with pulverized lead, zinc, or antimony, and vulcanizing the mass in the usual manner.

Cheap Erasing Rubber.

By dissolving caoutchouc in bisulphide of carbon until a uniform fluid has been obtained, and intimately mixing this with sufficient powdered starch to form a dough, and allowing the bisulphide of carbon to escape by placing the mixture in the open air, an elastic whitish mass is obtained which has the same effect upon pencil marks as pure rubber, but is a great deal cheaper. Finely powdered pumice stone may be added to the starch, which still more increases its erasing power. There can be no doubt that this mass can be vulcanized.

GUTTA PERCHA.

In the introduction of this work we have already given the most important items in regard to the manner in which gutta percha became known. In most all rubber factories it is worked in connection with caoutchouc, and mixtures of both substances are used for many industrial purposes. As far as its origin is concerned, we will state that it is also the dried milky juice of a plant, and that all the gutta percha which is found in commerce is obtained from one genus.

This plant, the *inosandra gutta*, belongs to the family of the *sapotaceæ*, and is an immense tree whose trunk reaches a height of 24 metres (78.72 feet), and a diameter of 2 metres (6.56 feet). The tree abounds over a vast extent of territory which includes, as far as is known at the present time, the southern part of the East Indies and the large islands of the Asiatic Archipelago.

Gutta percha is also a body which is found in solution in the milky juice of the tree. The vessels containing the milky juice run vertically between the bark and the wood of the tree, and can be easily recognized by dark lines running parallel with them on the surface of the bark. The milky juice is now obtained by making incisions in the trunk, but when gutta percha first became known and a sudden strong demand for it sprung up, the trees were simply felled, and in this manner, it is claimed, over 300,000 trees were destroyed in a few years.

While the milky juice of the caoutchouc trees remains fluid for a long time, that of the gutta percha tree congeals within a few minutes after it has been tapped, and this is said to be the case also if the juice is immediately put into bottles. Its behavior in this respect resembles that of blood drawn from a vein, it separates into a watery and into a solid part, the latter forming the substance which has become known by the name of gutta percha.

At the present time it is obtained by making an incision in the trunk and catching the juice in vessels. As soon as the juice has become sufficiently thick that it can be worked with the hands, the parts which remain fluid are separated from it by kneading. Usually the masses of gutta percha are formed into cakes about 30 centimetres (11.7 inches) long and 10 to 12 centimetres (3.9 to 4.6 inches) thick. In some localities it is gained in the following manner. The juice is boiled down and the residuum is freed from the fluid still adhering to it by kneading it with the hands.

XIV.

PROPERTIES OF GUTTA PERCHA.

The remarkable physical and chemical properties of gutta percha make it an object of the utmost importance, and, when the cost of the crude material is lessened, its use will be greatly extended.

A.—Physical Properties.

Commercial gutta percha represents differently colored, fibrous masses. The best varieties are nearly white or grayish-white, while others have a yellowish, reddish, or brown hue. This coloring very likely comes from changed juice which has not been entirely removed in preparing the article; pure gutta percha being a white body.

The commercial article is frequently mixed with pieces of wood and bark, and sometimes small stones are found in the interior of the cakes. To the touch it feels somewhat like the bark of a tree and possesses a peculiar odor resembling somewhat that of caoutchouc blended with a leather smell.

Crude commercial gutta percha, as well as the refined, floats upon water. This would indicate that gutta percha had less specific gravity than water; but this is not the case, for thin plates of it laid upon water and placed under the receiver of an air pump, sink as soon as the air is exhausted, the numerous small pores having absorbed the water.[1]

Even if only slightly cleansed it becomes so dense that no fluid can penetrate it, and the more it is cleansed the more compact and homogeneous does it become. If stored for any length of time it suffers an essential change through atmospheric influences and becomes friable and brittle. It can be improved by heating and kneading, but it is impossible to manufacture fine articles from such old gutta percha. The manufacturer should therefore try to obtain it in as fresh a condition as possible and to protect it against atmospheric influences. The simplest way of accomplishing this is by covering the gutta percha with water.

When exposed to the air it undergoes a process of oxidation, in consequence of which it becomes heavier and is transformed into a resinous body which possesses less hardness and can be easily dissolved in spirit of wine. Commercial gutta percha contains about 15 per cent. of this resin.

The behavior of gutta percha at different temperatures is of special importance. At an ordinary temperature its behavior is similar to that of wood or hard leather, only thin pieces can be bent by exerting con-

---

[1] *Payen,* on stretching gutta percha under strong pressure and immediately cutting the strips thus produced into very small pieces under water, found that the greater part of the fragments fell to the bottom of the vessel, some immediately, others after absorbing a certain quantity of water.—*Translator.*

siderable force. Its pliability quickly increases between 25° and 30° C. (77° and 86° F.) and if exposed to a temperature of 50° C. (122° F.) it becomes so soft that by using a powerful pressure it can be rolled out into plates. If it is heated to a temperature between 55° and 60° C. (131° and 140° F.) a mass is formed which, as far as plasticity is concerned, surpasses every other substance. If carefully heated to 120° C. (248° F.) it melts to a thin fluid, and, if heated still more, decomposition takes place, and among the products of dry distillation, the same bodies are found which can be obtained from caoutchouc.

It is inflammable at a certain temperature and burns with a bright flame, leaving behind a very small quantity of ashes. In contrast with caoutchouc, which seems to have no structure whatever, gutta percha possesses a fibrous structure. A slip cut from a thin sheet may be stretched considerably in one direction, that is, in a line with the fibres, but any attempt to stretch across this line is followed at once by a rupture. A sheet of caoutchouc, on the contrary, will stretch equally well in all directions.

Examined in thin sections under the microscope it seems to possess a porous structure. Caoutchouc under these conditions gives little or no change of color, while gutta percha exhibits a beautiful spectacle. It appears to be built up of prisms of every variety of hue, and, as it were, fused into each other. *Prof. Page* states that it resembles more nearly some specimens of ice which he has examined than anything else. The porous structure may be seen by allowing a drop of solution in bisulphide of carbon to evaporate spontaneously on a glass slide. The solution is soon reduced to a whitish plate, and the numerous cavities with which it is pierced may be distinctly perceived. The cavities may be made more visible by means of a drop of water; the liquid gradually insinuates itself, the mass appears more opaque, and the cavities are seen to be enlarged.

Gutta percha can be dissolved in all solvents which will dissolve caoutchouc, with this difference, that it is more easily soluble than the latter. It can be partly dissolved in anhydrous alcohol, and completely in ether, if it has not been previously treated with alcohol.

B.—Chemical Properties.

According to analysis gutta percha consists of a combination of hydrocarbons, the composition of which resembles very much those of caoutchouc. The oxygen which is found in it very likely belongs to a foreign combination. Gutta percha contains:—

Carbon	86.36
Hydrogen	12.15
Oxygen	1.49
	100.00

As far as the chemical composition of gutta percha is concerned it must not be considered as a single combination, but as a mixture of several bodies which occur in different quantities in different varieties of it. Two of these combinations can be dissolved by boiling in anhydrous spirit of wine, and if the solution is allowed to stand quietly for some time, small white grains are separated, the surface of which consists of numerous small crystals, but contain a yellow, amorphous core in the centre.

The yellow non-crystalline mass can be more easily dissolved in cold spirit of wine than the crystals, and according to this, gutta percha, by a suitable treatment, can be separated into three parts, one of which (gutta) is insoluble in alcohol, the next (albane) is difficult to dissolve, while the third (fluavile) is easily dissolved.

Payen, who thoroughly examined gutta percha, found in 100 parts of it:—

	Per cent.
Gutta	78 to 82
Albane	16 to 14
Fluavile	6 to 4

Pure *gutta*, which remains after the crude gutta percha has been thoroughly exhausted with alcohol, is a white mass, which, when rolled into thin sheets, is tenacious and ductile at a temperature of from 15° to 30° C. (59° to 86° F.), but not very elastic. It becomes soft when heated to 45° C. (113° F.) and assumes a yellowish color; the higher the temperature the darker and more transparent becomes the mass, and finally becomes dough-like but without actually melting. It melts when the temperature is raised to 120° C. (248° F.) and commences to decompose at a still higher temperature. Towards solvents its behavior is the same as that of gutta percha.

Albane is the crystalline resin which is separated from the solution in boiling spirit of wine. It melts at from 175° to 180° C. (347° to 356° F.). If exposed to a still greater heat it is decomposed and furnishes the same products as gutta.

Fluavile is a non-crystalline resin of an orange color. It is hard at an ordinary temperature, but becomes soft when taken in the hand, melts between 100° and 110° C. (212° and 230° F.) and is decomposed, when heated still more, emitting at the same time pungent vapors.

Recent investigations have clearly established the relation between gutta, albane and fluavile. According to these investigations we may suppose, that the body which is of actual value to us, namely, the pure gutta, is a hydrocarbon, being composed, according to *Baumhauer*, of $C_{20}H_{32}$. If we compare the formula for the composition of pure caoutchouc (C_5H_8), established by *Williams*, with that of gutta, we see that the formula of the latter is equal to four times of that of the first, and that, therefore, the two bodies show great similarity in regard to their chemical constitution.

Besides gutta ($C_{20}H_{32}$), we find, according to *Baumhauer's* investigation, two more combinations in the crude gutta percha. These are composed of $C_{20}H_{32}O$ and $C_{20}H_{32}O_2$, and are, therefore, products of oxidation of gutta. We might suppose that these products were already present in the fresh milky juice itself, but from the fact that, if gutta percha is stored for any length of time, its properties undergo an essential change, we must conclude that new products of oxidation are constantly formed.

Mr. Clark undertook a series of experiments upon gutta percha, and we give below the result as interpreted by *W. A. Miller.*

500 grains of a thin sheet of gutta percha were exposed for eight months under the following conditions:—

1. In netting open to the air and light, but excluded from rain.
2. In a bottle open to the air and light, but excluded from rain.
3. In a bottle open to the air, but excluded from light.
4. In fresh water, open to air and light.
5. In fresh water, open to air but excluded from light.
6. In fresh water, excluded from air and light.
7. In sea-water, exposed to air and light.
8. In sea-water, excluded from light but exposed to air.
9. In sea-water, excluded from light and air.

The specimens 4, 5, 6, 7, 8, and 9 were wholly unaltered, with the exception of a slight increase in weight, due to the absorption of water, which they lost again after exposure to the air for one or two hours. The tenacity and structure of the material did not appear to have undergone the slightest change.

No. 2, which had been folded up and introduced into a bottle, the mouth of which was open and inverted, had absorbed 5 per cent. of oxygen, 55 per cent. of the mass being converted into resin. The outer layers, exposed to light, were brittle and resinous, but the inner portions, screened from light by the outer folds, were but little altered in texture or appearance.

No. 3 had experienced little or no change, had increased in weight only 0.5 per cent., and yielded to alcohol only 7.4 per cent. of resinous matter.

Another sample which had been exposed to the light of day for a period of only two months, had become quite brittle, had increased in weight 3.6 per cent., and yielded 21.5 per cent. of resinous matter to alcohol, while a piece of the same sheet, kept in the dark, had undergone no sensible change.

Miller also examined several specimens of cables which had been submerged for periods of time varying from a few weeks to seven years. *In no case where the cable had been completely and continuously submerged, did he find any sensible deterioration in the quality of the gutta percha.* The only perceptible chemical difference in the various specimens was in the quantity of water mechanically retained in each.

One of the principal properties of gutta percha is its great chemical indifference. Concentrated alkaline solutions, as well as not too concentrated acids, and all salt solutions, do not affect it in the least. It only commences to char when subjected for a long time to the influence of concentrated sulphuric acid; but smoking sulphuric acid brings about a quicker change, and transforms it into a slimy substance.

Even the strongest hydrochloric acid seems to have but a very slow effect upon it; the only effect observed, after lying for months in the acid, was that it had lost some of its pliability. But cold concentrated nitric acid acts very energetically upon it, and, when boiled, dissolves it completely under the emission of red vapors.

Its chemical indifference and great plasticity make this body absolutely invaluable for certain branches of chemical industry. As it is entirely indifferent towards hydrochloric acid, it is used for manufacturing hose for drawing the acid from the vessels containing it, and is even employed for lining boxes in which the acid is to be transported.

It is also indifferent towards not too concentrated hydrofluoric acid, and is, therefore, employed for manufacturing vats which are to be used in etching glass, and, also, bottles in which the acid is to be kept.

XV.

CLEANSING OF CRUDE GUTTA PERCHA.

Crude commercial gutta percha is never pure enough to allow of it being worked without being subjected to a preparatory cleansing. The commercial article contains, besides pieces of bark and wood, such considerable quantities of earth and sand as to leave no doubt that they have been admixed with dishonest intent during the kneading of the hardening milky juice.

Before commencing the treatment of the raw material, it is best to examine its condition more accurately. This is done by taking a few cakes from the mass to be tested at random, and cutting them up with knives; if, besides gutta percha, only pieces of wood and earth are found, the mass can be at once brought into the cutting machines; but, if it contains stones, it is absolutely necessary that it should be subjected to a special operation before it is cut up, as, if this precaution is neglected, not only the cutting machines might be ruined, but accidents might also happen.

For the purpose of removing stones, the cakes are softened in water having a temperature of about 50° C. (122° F.), and are then rolled into thin bands. So that the stones may not stop or injure the rollers, they are so arranged that the upper one runs in a movable journal box which is held in place by a lever.

In case a stone gets between the rollers, the upper one lifts up and falls down again as soon as the stone has passed through.

By forming the gutta percha into bands in the above-described manner every stone can be immediately detected and removed. The bands are folded together, while still warm, in such a manner as to form loose blocks of a size corresponding to the cutting machines. The object of these machines is to transform the gutta percha into very fine shavings; and this is easier accomplished than is the case with caoutchouc, as the elasticity of the latter body makes this operation a rather difficult one.

The same machines which are used for cutting up caoutchouc may also be employed for gutta-percha; but, as a general rule, machines of special construction are used. Among these the drum slicing machine and the wheel slicing machine deserve special mention.

The drum slicing machine consists of two circular disks made to revolve as quickly as possible by a pulley, and are connected with each other by a large number of obliquely set knives placed upon the surface of the cylinder. A vertical cylindrical pipe is placed underneath the cutting machine, and in this is fitted a piston which is pressed upwards by a lever suitably weighted.

A cube of gutta percha is placed in the vertical pipe and pushed by the piston against the knives of the revolving disk and cut into thin slices. When the block has been cut down to a piece the pipe must be charged anew. This is done by cementing the remaining piece to a block which has been previously warmed, and the cutting is then continued.

The wheel cutting machines, as frequently used at this date, resemble straw cutting machines. They have a fly wheel about 2 metres (6.56 feet) in diameter with the cutting knives fastened on its spokes. The wheel makes from 500 to 600 revolutions per minute. The block of gutta percha placed upon an inclined plane is fed to the knives and cut into very thin shavings.

The shavings obtained in this manner are subjected to the actual cleansing process. According to the old method they were placed in a vessel filled with water, which was kept in constant circulation by a stirring apparatus. The gutta percha would float on the surface while the pebbles and earth fell to the bottom. The washed gutta percha was then placed in a vessel filled with warm water and pressed into a band by being passed through rollers. This was again passed through a second pair of rollers rotating somewhat quicker than the first pair, and stretched while passing through them. Finally, by using a sufficient number of rollers, 5 to 6 pair, a long, thin band of cleansed gutta percha was obtained. Although this method is a very practicable one, it possesses the disadvantage of requiring not only a very complicated apparatus, but also considerable mechanical power. A somewhat different method for obtaining the gutta percha in a cleansed condition and at the same time in the form of lumps is now employed in the largest factories.

The shavings are brought into a large vessel filled with water into which steam is admitted. When the water has been brought to the boiling point it is thoroughly stirred. By doing this the shavings ball together and form large lumps, and these are thrown into the so-called tear-wolf (zereiss-wolf).

This apparatus consists of an iron cylinder the entire surface of which is covered with curved, iron teeth. It is surrounded by a drum, the space between this and the teeth being very narrow. The soft masses of gutta percha are pressed against the drum, which revolves from 700 to 800 times in a minute, a stream of water being introduced at the same time. This apparatus tears the gutta percha into extremely fine shreds and removes all impurities from it.

These shreds are softened in hot water and brought to the kneading machine in which they are combined into balls or lumps. The kneading machine consists of a strong, hollow cylinder heated by steam by outside connecting pipes. Inside the cylinder are four to six rollers covered with blunt tapering teeth. The rollers press the mass against the inner surface of the cylinder, the teeth expelling all the air and water contained in it. The thoroughly kneaded mass, while warm, possesses extraordinary plasticity, and can be immediately used for fashioning various articles.

As may be seen from the foregoing description a

number of apparatuses and considerable mechanical power are required for cleansing gutta percha. To avoid this, experiments in cleansing gutta percha by use of a solvent have been made, but not introduced into general use, though the method has some advantages. The simplest plan would be to wash and dry the shavings and dissolve them in bisulphide of carbon. The solution is allowed to clear by settling and the solvent is distilled off in a suitable vessel, the purified gutta percha remaining behind in a homogeneous mass.

At first glance this process seems a very simple one, yet it has many difficulties, the greatest one being to remove all traces of the solvent, but this can be accomplished by heating the mass after distillation to 110° or 115° C. (230° to 239° F.).

XVI.

VULCANIZATION OF GUTTA PERCHA.

The internal properties of gutta percha, as has been shown, closely resemble those of caoutchouc, and it can be subjected to vulcanization, the method being almost the same as for coautchouc. The gutta percha is mixed either with sulphur or sulphides, or it is treated with chloride of sulphur and then burned.

Vulcanized gutta percha retains its tenacity and pliability in all changes of temperature, and acquires a still greater power of resisting chemical influences than the pure product.

In vulcanizing it with pure sulphur one point must always be remembered, namely, that a much smaller quantity of sulphur is required for vulcanizing gutta percha than for caoutchouc, and that a brittle product will be the result if an excess of sulphur is used. Pure sulphur alone is seldom used for vulcanizing it, either sulphur in connection with sulpho-metals or chloride of sulphur being employed for the purpose.

Practical experience has shown that it is best to use sulpho-metals in connection with a small quantity of sulphur. According to an English receipt, the following mixture is used for preparing vulcanite of gutta percha:—

	Parts by weight.
Gutta percha	48
Sulphur	1
Sulphide of antimony	6

The substances are mixed in a manner similar to that which has been described for preparing caoutchouc vulcanite. The gutta percha is burned at a temperature of from 125° to 150° C. (257° to 302° F.).

The great plasticity gutta percha acquires at a higher temperature makes it possible to incorporate solid foreign substances with it, masses being formed in this manner, which differ very much in character. Later on we will discuss this matter more fully.

The simplest plan to vulcanize gutta percha is to dissolve it in bisulphide of carbon to the consistency of syrup, and mix it into a solution of chloride of sulphur. If too much of the last solution is used, that is to say, if the quantity of chloride of sulphur amounts to 10 per cent. of the weight of the gutta percha, a mass is obtained which does not become soft even if exposed to a temperature of 100° C. (212° F.). The hardness increases in proportion to the quantity of chloride of sulphur, and by using about 15 per cent. of the latter, a mass resembling horn, as far as its properties are concerned, will be the result.

A thick solution in bisulphide of carbon is used for coating articles with vulcanized gutta percha. The articles are brushed over with the solution, and, when dry on the surface, are plunged into a solution containing 5 to 10 per cent. of chloride of sulphur to 100 parts of bisulphide of carbon.

XVII.

BLEACHING OF GUTTA PERCHA.

GUTTA PERCHA, like caoutchouc, remains unchanged in its physical and chemical properties by bleaching. Its great indifference towards chemical agents, and its faculty of being colored as to resemble human gums so as to defy detection, make it of immense value to dentists.

There are two bleaching processes, namely, by using chloroform or animal charcoal.

If chloroform is to be used, the commercial gutta percha is cut up into small pieces. Twenty times the quantity of chloroform is poured over these, and when all has been dissolved, which requires from three to four days, a small quantity of water (about $\frac{1}{4}$ litre (0.44 pint)) is added to the solution. The vessel containing the mass is then thoroughly shaken and allowed to stand quietly until the contents have become separated into two distinct layers. The lower of these layers consists of a solution of pure gutta percha in chloroform, while the one floating on top consists of the water and the foreign substances which had been mixed with the gutta percha.

The clear solution is then drawn off by means of a siphon into a porcelain basin. This is placed in a copper still and surrounded with water. The still is then closed and heated, and all the chloroform is distilled off.

The purified gutta percha remains in the porcelain basin in the form of a vesicular mass, which can be formed into plates and small sticks by softening it in hot water and by mechanical treatment. While a purified article is obtained in this manner, it is not entirely decolorized, as it always shows a weak yellowish or brownish color.

To entirely decolorize it and to obtain a pure white mass, we propose the following process: Purified gutta percha is dissolved in twenty times the quantity of bisulphide of carbon. The solution is clarified by allowing it to stand quietly, and is then filtered through finely powdered animal charcoal. But on account of the great volatility of the solvent, it is necessary to use a suitable apparatus for filtering.

An apparatus of this kind, of simple construction, and performing excellent service, is represented by the accompanying illustration (Fig. 5). It consists of a

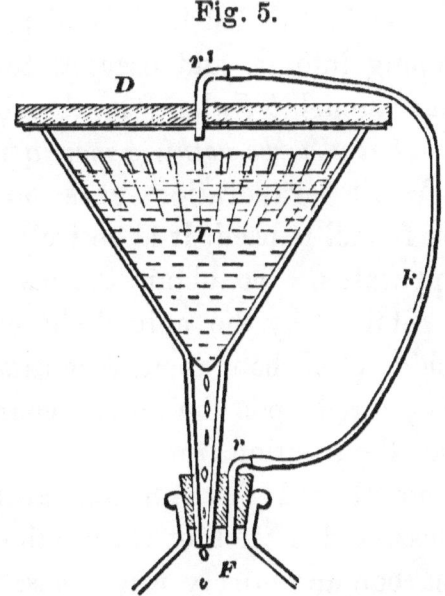

Fig. 5.

large bottle, *F*, either of glass or tin. This is hermetically closed by a cork with two holes. The neck of the glass funnel, *T*, the upper rim of which is ground smooth, is placed in one of the holes, and a glass tube, *r*, bent at a right angle, is fitted into the second hole. A thick wooden lid, with a ring of rubber on the lower side, is placed upon the funnel, thus closing it airtight. In the centre of the lid is fitted a glass tube, r', also bent at a right angle, which is connected with the tube r by a rubber hose, k.

The funnel, the neck of which is closed by a stopper of cotton, is filled about two-thirds full with animal charcoal. The solution to be filtered is poured upon this, the lid is placed in position, and must be removed only for the purpose of pouring more solution into the funnel. The air in the bottle F is displaced by the solution dropping into it, and escapes through r, k, and r' into the funnel T, where it absorbs the vapor of the solution, *but absorbs nothing more after it is once saturated.* While evaporation goes on constantly when an open funnel is used, it is entirely checked by using this apparatus. To obtain the last remnant of the solution retained by the animal charcoal, a quantity of bisulphide of carbon, about 2 centimetres (0.78 inch) high, is poured upon the animal charcoal, which will remove all the solution from it.

The solution filtered through animal charcoal is almost colorless, and after the evaporation of the bisulphide of carbon an entirely white mass is obtained which can be colored with the most delicate coloring matter. To remove the last traces of the solvent the mass is heated for some time at 100° C. (212° F.).

Solutions of gutta percha bleached in this manner are entirely colorless, and when spread upon glass plates furnish a coating resembling a film of collodion, but have the advantage of possessing greater solidity and tenacity. As the preparation of bleached gutta percha is rather complicated, it is scarcely employed for any other but dental purposes, although an excellent use could be made of it for ivory compositions.

XVIII.

GUTTA PERCHA COMPOSITIONS.

GUTTA PERCHA has the same capacity as caoutchouc of combining, when softened by heat, with a variety of substances. With suitable components, compositions can be made resembling leather, wood, whalebone, horn, and even stone.

Soon after gutta percha became known, the hope arose of making from it a substitute for leather, and efforts made in that direction have been so far successful as to prepare compositions which serve many purposes formerly supplied by leather, but not quite equal in tenacity and solidity to a well tanned leather.

Compositions of gutta percha and caoutchouc are of great importance in the preparation of matrices for the rapidly developing industry of electrotyping.

Gutta percha, by itself, can be very well employed to receive imprints from coins, or medal dies, by merely heating and subjecting it to a high pressure under the die until it becomes cold. The imprint on the gutta percha is the negative of the coin or medal in all of its finest details and preferable to plaster of

Paris impressions, as they can be repeatedly used for making electrotype copies.

But compositions of gutta percha and caoutchouc must be used for overlaying matrices or moulds, to unite plasticity, when heated, with sufficient elasticity to allow of the matrix being removed without injury to the impression.

The proportion required for the caoutchouc and gutta percha composition best adapted for this purpose can only be ascertained by special experiments. A more elastic (richer in caoutchouc) mixture must be used for deeply cut models, which are to be copied (high reliefs) than is necessary for copying a less projecting article (low relief). The best plan for mixing caoutchouc and gutta percha into a homogeneous mass is to pass plates of the two bodies through heated rollers revolving at unequal velocities, to cut up or fold up the plate thus formed, to again pass this through the rollers and to repeat this operation until an entirely homogeneous mass has been formed.

Gutta Percha and Caoutchouc Composition for Machine Belts.

Compositions of caoutchouc and gutta percha prepared in suitable proportions combine great tenacity and solidity with a certain degree of elasticity, and can therefore be advantageously used for the manufacture of machine belts. Although their first cost is a considerable one, they are cheaper in the end, as they are almost indestructible, and besides can be easily repaired. A composition answering very well for this purpose consists of:—

	Parts by weight.
Gutta percha	70 to 75
Caoutchouc	30 to 25
Sulphide of antimony	5 to 4
Sulphur	2 to 1

To obtain as intimate a mixture as possible, it is advisable to weigh the caoutchouc and gutta percha in the form of shavings, to mix them thoroughly together and form bands in the manner already described. These bands should be rolled until they are entirely homogeneous, the sulphur and sulphide of antimony being incorporated with them at the same time.

The mass, when finished, is formed into a block, and this is passed through between the rollers at a rather low temperature, and is gradually changed into a band corresponding in width with that of the belt to be manufactured. When the thickness of this band nearly approaches that which the belt is to have, the temperature is lowered so much that the band can be forced through between the rollers only at the expense of great power, to make the mass as compact as possible.

The edges of the belt are then trimmed, and it is covered on one side with a linen or cotton cloth, and wrapped loosely around a wooden roller in such a manner that the cloth forms a layer separating the windings of the belt from each other. (If this were neglected the mass would fuse together during the burning process.) It is then subjected to the burning process during which the temperature, especially for thick belts, must be raised to 160° C. (320° F.). When the belt has been sufficiently burned, it is taken from the wooden roller, smoothed, and polished by being passed through between smooth rollers (calendering).

Hard Gutta Percha Compositions.

Various admixtures are used for hard gutta percha compositions, and are nearly the same as mentioned for caoutchouc, including earths, oxides, and finely pulverized minerals. While some of the ingredients are selected to give weight or color to the mass, many have no other use except economy in making the necessary bulk.

Whiting, white pipe-clay, magnesia, oxide of zinc, washed barytes or artificially prepared sulphate of baryta are used for white, or rather yellowish colored articles. If they are to be of light weight, it is advisable to use magnesia; for heavy objects, sulphate of baryta is the best.

On account of the brown color of unbleached gutta percha, the composition prepared with white materials will never show an entirely white, but always a more or less yellowish-brown color. If it is desired to obtain entirely white compositions, it will be necessary to use bleached gutta percha.

GUTTA PERCHA COMPOSITIONS. 175

The very cheap colcothar (caput mortuum) can be used as a component for reddish-brown compositions; powdered pyrolusite for dark-brown masses; and black compositions may be prepared by incorporating bone-black, etc., with the gutta percha.

The total weight of the added ingredients may be greater than that of the original gutta percha, without destroying its plastic property, or preventing its being pressed into any desired shape. But such compositions will always show some degree of brittleness, and for this reason their use is limited to such articles as are not to be exposed to shocks, as door handles, escutcheons, ornaments for frames, etc.

For articles exposed to shocks, the foreign admixtures must not amount to more than from 25 to 30 per cent. of the weight of the gutta percha, and from such masses may be manufactured many of the utensils for daily use which were formerly constructed of leather, tin, or wood.

To hide the peculiar and not agreeable odor of gutta percha, sweet smelling substances should be mixed with it; essential oils have frequently been used for the purpose. While these completely disguise the odor of gutta percha, they possess the disadvantage of gradually volatilizing. It is, therefore, advisable to choose substances which will retain their perfume for a long time; benzoin, tonka beans, and orris root can be recommended. Of benzoin, 4 per cent. of the weight of the mass is sufficient to make the odor an agreeable one; of tonka beans $\frac{1}{2}$ per cent. is more than sufficient; if orris root (the hard root of *Iris florentinæ*) is used, about 10 per cent. will be required, as its perfume is not very strong. Fine shavings of sandal wood or of American juniper (*Juniperus virginiana*) may be used instead of orris root.

Compositions of Gutta Percha and Wood.

For many years, admixtures of finely pulverized cocoa-nut shells have been used in gutta percha compositions, giving to them the properties of wood. The shells are powdered in stamp mills; the powder is sifted and incorporated with the gutta percha in the usual manner.

But powdered wood and sawdust of hard woods can also be incorporated into these compositions. The sawdust is ground and sifted into a fine flour. But it is absolutely necessary to thoroughly dry the powdered wood before mixing it with the gutta percha, and it may be further recommended to coat the plates, made from such compositions, with a gutta percha varnish.

Compositions of gutta percha and wood can, like wood, be worked by means of the saw and turning lathe, and can be very advantageously used for covering wooden articles. But for the wood and composition to adhere tightly to each other it becomes necessary to frequently saturate the first with linseed oil, before the composition is laid on, to prevent it from absorbing atmospheric moisture. If this precaution is neglected and the wood should become damp, the composition would fall off in consequence of the great expansion of the wood.

By coating two boards saturated with linseed oil with a gutta percha wood composition, placing them cross grained upon each other and heating sufficiently to soften the gutta percha, and then submitting them to a heavy pressure until cold, a tablet of extraordinary tenacity and resistance, exceeding those of the hardest woods will be formed.

Sorel's Gutta Percha Compositions.

The so-called substitutes for caoutchouc and gutta percha, which were introduced by *Sorel*, should actually be considered as gutta percha compositions in which the latter is mixed with pitch, resin, lime, and potter's clay.

The best of *Sorel's* compositions consist of:—

	Parts.
Colophony	2
Pitch or asphaltum	2
Rosin-oil	8
Slacked lime	6
Water	3
Potter's clay	10
Gutta percha	12

The purpose of the rosin-oil in this composition is evidently to dissolve the pitch and colophony, and the admixture of lime is very likely for the purpose of effecting a combination between the acids of the colophony and the lime. The potter's clay is introduced into the composition as an indifferent substance to increase the weight of the mass, and can, therefore, be replaced by other indifferent substances (chalk, magnesia, colcothar, etc.), or can be entirely left out. In the latter case a composition is obtained which, in regard to its properties, approaches that of pure gutta percha.

The manner of preparing the composition is as follows: The colophony, pitch, and rosin-oil are stirred together in a boiler until a complete solution has been accomplished, which can be accelerated by heating the substances. The lime and water mixed together to a paste are then added, and finally the gutta percha and potter's clay, but the latter only when the gutta percha has become fluid. More water is added to the mass, and it is heated to the boiling point of water (100° C., 212° F.), and then taken from the boiler.

Even if the work is carried on with the greatest care, it will be impossible to obtain in this manner an entirely homogeneous composition. To do this, it is necessary to pass the mass several times through between the rollers. 5 per cent. of stearic acid or wax should be added to the composition for the purpose of making it entirely water proof.

Sorel varies his mixtures to suit different purposes and claims that they can be substituted for pure gutta percha; but his compositions have not the great tenacity of the pure material, nor its indifference to chemical action, and cannot be advantageously used for articles liable to exposures to chemical agents. The following are some of *Sorel's* receipts for manufacturing gutta percha compositions:—

	I. Parts.	II. Parts.	III. Parts.
Pitch	8	12	—
Rosin-oil	4	—	—
Coal tar	—	—	12
Slacked lime	6	6	6
Gutta percha	16	16	16

ROUSSEAU'S METHOD OF PREPARING SOLUTIONS OF GUTTA PERCHA AND THEIR USE.

Gutta percha, to which linseed oil has been added, is heated in a suitable vessel over an open fire. Linseed oil generally absorbs one-tenth of its weight of gutta percha.

When a white cotton fabric is dipped into this solution it will be thoroughly penetrated by it, and, after it has become cold, will be of a yellowish color, transparent, and very soft. Such material can afterwards be printed with all kinds of colors.

Whiting, ochre, lampblack, etc., may be added to the solution for the purpose of thickening and coloring it, which will also remove the peculiar odor of the gutta percha.

For lacquering leather, coating of taffeta and gauze, some copal varnish, to which the gutta percha imparts its softness and elasticity, must be mixed with the solution.

It can be mixed with all substances—gutta percha especially exerting no influence whatever upon oil paints.

XIX.

MANNER OF WORKING CAOUTCHOUC AND GUTTA PERCHA.

Since the invention of the process of converting caoutchouc by vulcanization into vulcanite and hard rubber, and the introduction of gutta percha as a new material possessing certain properties lacking in caoutchouc, the use of these bodies, and of articles manufactured from them, has been extended in an extraordinary manner.

And yet the working up of these peculiar products of nature is still in its infancy. As it can be rolled into sheets of the greatest thinness, it seems to be destined to replace paper in various respects. Maps, globes, etc., are already prepared from it. An extraneous circumstance will promote this development. It is already regarded as a fact that the consumption of paper is now out of proportion to the production of the raw material necessary for it, to wit, rags. All efforts to check the increase of this disproportion, through the use of other raw materials, have as yet produced but an incomplete—by no means effective—result. This is particularly noticeable in the United States, which, as of so many other things, can boast of a grand journalistic and other literary productiveness, and vainly looks for raw materials in the markets of the world for its immense paper consumption. The remedy will not long be sought for; the indications are already given. Bleached gutta percha, especially, is better adapted for lithographic printing than the finest Chinese paper, it yielding really admirable copies. By wetting it with a solution of gutta percha in sulphuretted carbon, printing paper can most easily be transformed into writing paper.

The American division of the Paris Exhibition owed its principal attraction to its numberless India-rubber articles. People gazed there with admiration at handles of knives and rifle-stocks adorned with the finest and most artistic reliefs, at opera-glasses, and a thousand other optical instruments, or articles of cabinet-ware, which were formerly manufactured of ebony or buffalo's horn. There were also exhibited richly gilt pieces of furniture wrought entirely of this material, as well as articles of vertu set with genuine pearls, and various utensils ornamented with Chinese paintings. We observed, further, musical instruments, such as violins, clarionets, and trumpets. Whenever we visited the exhibition, we could not refrain from admiring the music executed on one of those trumpets shortly before the close, which was indicated by the ringing of all the bells contained in the building. To make the contents of the whole collection more complete, we must add candelabras, electric machines, very flexible whips and canes, surgical instruments of every kind, powder-horns, printing type, spools, shuttles, slates. Further, articles of caoutchouc and rubber compositions representing columns, pillars, small monuments, etc., which, in regard to their external appearance, could scarcely be distinguished from the most beautiful marble, porphyry, serpentine, and other expensive stones.

Neither was the art of printing forgotten in this rich collection. A thick quarto volume contained the history of the industry. The leaves challenged destruction by water, being made of vulcanized caoutchouc, as also the elegant binding. It is greatly to be regretted that this invention was made at so late a period; made earlier, it might have saved us many treasures of antiquity. The deluge itself would have been powerless to destroy such written monuments.

The chaotic variety of articles manufactured of vulcanized caoutchouc and gutta-percha is almost bewildering. Elastic textures of every description (silk, linen, cotton), elastic stockings for persons suffering with the gout, gaiters, garters, suspenders, drawers; elastic bands, cords, belts, telegraph wires; aprons, window shades, carpets, gloves; stoppers, bungs, diving apparatus, life-boats, bathing tubs; mattresses, pillows, tents; numberless articles for hunters, fishermen, travellers and photographers; utensils for the preservation of acids, bottles of every kind, cases, balloons, doll-heads, spinning cards, hurdles, troughs, pumps, umbrellas; and a thousand other objects.

In the foregoing chapters we have described the manifold mechanical and chemical labors required in order to purify the crude commercial products (caoutchouc and gutta percha), and to change them into homogeneous bodies, and also the manner in which vulcanite and hard rubber can be prepared. We then described the methods for preparing caoutchouc and gutta percha compositions, the manner of dissolving these bodies, etc. It remains yet to describe the processes which enable us to manufacture various articles from caoutchouc, gutta percha, and compositions.

The limit of our work will not permit us to enter into all the details of manufacturing. We can only discuss the general principles, and must leave the overcoming of special difficulties to the inventive genius of the manufacturer.

We will limit our directions to the preparation from caoutchouc and gutta percha of plates, bands, threads, hose; of hollow articles, as balls, globes, etc.; of waterproof materials, insulated telegraph wires, rubber shoes, etc.

XX.

MANUFACTURE OF CAOUTCHOUC AND GUTTA PERCHA PLATES.

Caoutchouc and gutta percha are much used in their natural condition, as well as in form of vulcanite for making plates of any desired thickness. In fact, plates from the thickness of paper to that of several centimetres are manufactured, either by cutting them from blocks, by evaporation of solutions, or by rolling.

Cutting Plates from Blocks.

Cutting plates from blocks is only used when pure caoutchouc is worked and when very thin plates and bands are desired. Vulcanite and gutta percha are almost exclusively made into plates by rolling. The operation of evaporating solutions is only employed in special cases when exceedingly thin sheets are to be manufactured.

The manner of cutting plates of a certain size and thickness is as follows:—

The purified shavings of caoutchouc are pressed into a block having the same form (round, square, etc.) as it is desired to give to the plates. This block is placed upon a cutting machine which is so arranged that the block is pushed forward a certain distance each time after the knife has passed down; this regulates the thickness of the plate to be cut.

The knife for cutting the plates consists of a very thin and finely ground steel blade set into a guide and is moved to and fro with great velocity by a suitable mechanism. To obtain a smooth clean cut, it is necessary for the knife to be pushed back and forward 800 to 900 times in a minute.

On account of the toughness of caoutchouc and the great friction, the knife would become so hot that the cut plate would stick to it. To avoid this, and at the same time to decrease the friction, a jet of cold water is allowed to fall constantly upon the knife. If the machine is constructed in the right manner, and the knife moves with sufficient velocity, the cut surfaces of the caoutchouc plates should be entirely smooth, and it should not be possible to detect the separate small cuts the knife has made.

Cutting Plates from Cylinders.

A peculiar method is employed in cutting plates of any desired length and up to a few centimetres thick. This is done by cutting the surface of a cylinder of caoutchouc from the base upwards in the form of a spiral. This method is admirably adapted for cutting long strips all in one piece, either to be used as inclosures or to be cut into threads. But it must be observed that only cylinders of Para caoutchouc can be cut into plates; the East India variety does not possess the necessary solidity, and strips cut from it in this manner would tear.

The machine used for this purpose consists of two uprights upon which rest two disks in a horizontal position. The surfaces of the disks are provided with pointed pins for holding the cylinder of caoutchouc which is placed between them. On one of the disks is a cog-wheel connected with a motor, and gives a uniform rotation of the cylinder around its axis. The knife standing obliquely towards the cylinder and moving quickly to and fro is pressed against it by spiral springs fastened on the guide, the pressure being regulated by a lever.

The height of the cylinder to be cut into plates should be from 30 to 50 centimetres (11.8 to 19.6 inches) but its diameter will depend on the length of the strip to be cut. When finally the cylinder has been cut up, a smaller one with a diameter equal to that of the guide disks will remain. The following formula may be used for calculating in advance the length of a plate or strip which can be obtained from a cylinder of known diameter:—

$$L = 3.1416 \frac{D-d}{4} \frac{}{\delta}$$

D is the diameter of the cylinder to be cut, d the diameter of the small cylinder remaining uncut, δ the thickness of the plate to be cut, and L the length.

Beautiful plates can be obtained by the above-described method if the workman possesses sufficient skill, and caoutchouc of a suitable quality is used. The resulting plates may be cut into threads or used as inclosures between two tissues for the manufacture of water-proof materials.

Manufacture of Plates from Solutions.

Sollier's process can be used to prepare thick plates from caoutchouc solutions. He spreads linen cloth with several layers of paste as evenly as possible, and, in order to secure a perfectly smooth surface, coats over the paste with a mixture of glue and molasses, which will dry entirely smooth.

The sheet is then stretched in order to avoid any wrinkles, and covered with a caoutchouc paste of uniform consistency. When it has become perfectly dry, the plate can be removed from the linen sheet.

But this plan has the serious objection that the solvent used in preparing the solution is lost. It is true a portion of it may be regained by bringing the linen sheet coated with the paste into a chamber having a temperature of 45° C. (113° F.). The air of this chamber is pumped out and driven through very cold pipes in which the solvent is condensed.

Although it is possible to recover in this manner a part of the solvent, it must not be forgotten that the workmen, while applying the dough, are very much annoyed by the poisonous vapors of the bisulphide of carbon.

It is only advisable to use solutions when plates too thin to be cut or rolled are to be made. We prepare plates adapted for water-proof coverings, surgical purposes and manufacture of balloons, etc., by allowing a clear solution of caoutchouc to dry upon a smooth glass plate.

An entirely clear solution of caoutchouc, which has been filtered through cotton, is thinned until it is of a sufficient consistency to form a thin sheet or film. Tests should be made upon pieces of glass before the plates are prepared. When the film remaining upon the glass has the desired thickness, the solution is not reduced any further, and the preparation of plates is commenced.

A skilled workman can prepare a plate measuring one square metre (10.76 sq. feet) by poising a pane of glass of the necessary size upon the outstretched fingers of his left hand, and inclining the opposite diagonal corners, so that the solution poured upon the upper corner will flow down and cover the entire glass. When this is finished, the pane of glass is placed upon a horizontal surface until the solution has become entirely dry. A cut is then made along the edge of the glass to facilitate the removal of the caoutchouc plate. Should the solution have formed a thick rim on the edge of the glass, the cut must be made inside of the rim, and this must be removed before the caoutchouc plate can be drawn from the glass. Of course the size of the plate corresponds to that of the glass used. Skilled manipulation of a homogeneous, clear solution will give plates only slightly tinted. Beautifully colored plates can be prepared by adding dissolved aniline colors to the solution. Quite large balloons can be made from such plates by cutting sheets of the proper size and fastening them together by moistening the edges with the solution and placing them under a heavy pressure. These balloons will remain filled for a long time, as gas permeates very slowly through caoutchouc sheets.

Preparation of Plates by Rolling.

A great majority of manufacturers, at this date, prefer the rolling of plates to any other process, as the best for securing uniform density and thickness. Large and thick plates can scarcely be prepared in any other way. The rollers have been so much improved in their construction that they can be heated by steam and set at any distance from each other, permitting the manufacture of even the thinnest plates.

Calenders.

These machines, as just described, are called calenders, and consist generally of two or three pairs of rollers of the same diameter made of cast iron with finely polished surfaces. Their velocity of revolving is regulated by a cog wheel placed on the end of each shaft and geared to fit exactly into each other.

The rollers, as has been mentioned, are hollow and their journals can be shifted between the slits of two very strong iron housings. In calenders with four rollers, the upper roller of the lower set is permanent, and receives the power which is communicated by the cog-wheels to the other rollers. An endless screw in the housings serves to move the opposite journals so as to maintain all the rollers in parallel positions. With the use of four rollers, three distinct sizes of plates can be rolled.

The rollers are steam-heated by means of a pipe connected with each journal of the rollers, and a steam discharge pipe on the opposite journal.

Plates of any desired width can be prepared by fastening cleets upon the upper pair of rollers with distances apart to suit the plate, and thus plates nearly as wide as the rollers are long can be prepared, or, by placing the cleets closely together, bands but a few centimetres wide can be turned out.

Caoutchouc, as has been mentioned, loses its elasticity at 40° C. (104° F.). The blocks of it to be rolled into plates must therefore be heated to this degree, and the rollers must be at least equally warm. But it is better to heat them somewhat more than this, as at the above temperature considerable power is required to form the blocks into plates.

In working, the rollers are heated to 80° C. (176° F.), and sometimes even to 100° C. (212° F.), as only comparatively small power is then required for stretching the block of caoutchouc. But as it becomes quite soft at this heat, it will be necessary to cool it off as quickly as possible to remove all stickiness. The simplest way of doing this is to pass the plates as soon as they come from the rollers into a vessel filled with cold water, and then to roll them up. But the latter should only be done after the plates have been cooled off to an ordinary temperature, as if this precaution is not observed, they would stick together and form a mass which could not be separated.

To prevent any interruption, the rollers should be accurately set before the work commences. When the block has gone through between the lowest pair of rollers, it is at once passed through between the next pair, and this is continued until plates of the desired thickness have been obtained. By using well-constructed calenders and conducting the work in a proper manner, faultless plates of uniform thickness and great length can be prepared.

Stretching Machines.

For rolling gutta percha either the calenders used for rolling caoutchouc may be employed, or a stretching apparatus of very simple construction. This consists of two highly polished steel rollers lying vertically one above the other. A level polished steel plate is placed beneath the lower roller.

When gutta percha is to be changed into plates with this machine, a lump of it is first softened by heat, and then beaten with a wooden mallet into an oblong block running to a point, the end of which should be narrow enough to allow of it being pushed between the lower roller and the steel plate.

The lower roller is set in motion, and grasps the

block and presses it through between the roller and plate. The block is then returned between the first and second rollers, pressing it still thinner. It is then further reduced by returning it between the second and third roller, and moves in the same direction as it did between the roller and plate.

The stretching machines are so arranged that by means of screws the rollers can be set at any distance from each other, thus making it possible to prepare plates of any desired thickness. This machine can be used for gutta percha instead of the calenders, as heating the block before passing it through the rollers is sufficient to give it the necessary degree of softness.

The band coming from the stretching machine should be cooled off as quickly as possible before it is rolled up as, if this precaution is neglected, and the band remains only slightly warm, it will certainly stick together and form one mass.

Many contrivances are used for the above purpose, but the most suitable plan is to allow the plate or band as it comes from the stretching machine to glide upon an endless cloth, which should be as long as possible and move at the same rate as the rollers. A current of cold air is blown by a ventilator upon the plate or band by which it will be sufficiently cooled off to lose all stickiness.

Instead of the current of air it is still better to allow a spray of cold water to fall upon the plate or band.

By this it will be so thoroughly cooled off as to allow of its being immediately rolled up.

PREPARATION OF PLATES FROM VULCANITE MASSES.

One of the processes just described can be employed to prepare plates from pure caoutchouc or gutta percha, by passing them several times through between rollers which are graduated closer and closer until very thin plates are produced.

But the case is different when plates are to be manufactured from vulcanite masses. These, as we know, consist of a homogeneous compound of caoutchouc and sulphur, frequently combined with other ingredients to give them bulk. They lose much of their cohesive power by these admixtures, and it is only practicable to prepare plates, rarely exceeding 5 millimetres (0.19 inch) in thickness, by passing them once through between the rollers, as it is easier to burn them uniformly.

The finished plate must be treated with the greatest care, as it becomes very soft during the burning. Small plates may be slipped upon a smooth piece of zinc of a suitable size, and thus subjected to the burning process.

But a great many difficulties present themselves when larger and entirely smooth plates are to be prepared. These can only be overcome by passing a very thin, smooth piece of zinc through between the rollers simultaneously with the vulcanite mass, and carefully rolling the zinc, with the plate upon it, round a cylinder, the sheet of zinc between the windings of the plate, preventing the mass from sticking together during the burning process.

It is an easier matter to prepare large plates, if absolute smoothness is not required. In such a case a fine linen cloth, which has been previously moistened, is passed through between the rollers together with the vulcanite, and both are wound upon a cylinder. The face of the finished plate will have the appearance of a tissue as the elevations and depressions of the cloth have been pressed into the soft mass.

For certain purposes, indented or embossed rollers are used, which stamp the design upon the plates. If, for instance, plates for shoe-soles are to be manufactured, one roller is engraved in such a manner that the face of the plate will be covered with small corrugations, which prevent the wearer of such shoes from slipping.

The plates can of course be provided with any desired design, as it is only necessary to engrave the rollers accordingly. Door mats of vulcanized caoutchouc, and also the small blocks of vulcanite known as erasing rubber are manufactured in this manner.

Thick plates of vulcanite are generally prepared from thin plates, as in this manner they can be made more uniform. This is done by placing a thin plate upon a level table in such a manner that it bears everywhere equally well. Upon this is placed a smooth sheet of zinc upon which rests another plate.

But this sheet of zinc must be placed so that the edge of the plate upon the table projects somewhat. The plate resting upon the zinc is now pushed forward until it rests upon the plate on the table.

A workman then places a roller which is somewhat longer than the width of the plates, upon the end where they join, and applies a gentle pressure. While a second workman slowly draws away the sheet of zinc lying between the two plates, the first follows with the roller, and by applying a gentle pressure, unites the two plates into one. This labor requires considerable manual skill, as it is especially necessary to force out all the air from between the two plates. Should even the smallest air bubble be left, the place in the plate where it is, would swell up very much during the subsequent burning.

Should the plate not be thick enough, the above-described operation is repeated, and a third plate is laid upon the two already combined in one. Sometimes a caoutchouc mass mixed with sulphide of antimony is used for the centre plate, while the two outer plates consist of the ordinary mixture of caoutchouc and sulphur—a cross-section of such plates showing a reddish-brown stripe after they have been burned.

The ordinary solidity of vulcanite is not sufficient for thicker plates, to be used for machine packing. To obtain the desired solidity a linen cloth is spread upon a plate and covered with another plate, and the three pieces are then united by rolling.

XXI.

MODE OF PREPARING CAOUTCHOUC AND GUTTA PERCHA THREADS.

The fabrication of threads forms an important part in the working of caoutchouc, as, on account of their elasticity and tenacity, they are extensively used for the manufacture of elastic tissues. How extensive this use is may be judged from the fact that there are large factories entirely devoted to the manufacture of materials from which the elastic webbing for shoes is made.

Caoutchouc threads may be prepared according to various methods, but their elasticity and toughness depend on the raw material used. We will remark here that caoutchouc, which has been comminuted, masticated, and then brought into a compact mass by rolling, never possesses the solidity and elasticity of the raw material, and, of course, threads manufactured from the former are of a poorer quality than those prepared from the latter.

Square Cords from Crude Caoutchouc.

The best quality of caoutchouc, in the form of bottles with the thickest sides and most regular shape, is always selected for making such cords. The necks of the bottles are cut off and the bowl or body is bisected by a cross-cut. The pieces are then examined, and only those having an entirely homogeneous appearance are selected.

The next step is to convert the pieces of the bottle into plates. To do this they are softened by a continued boiling in water, and then placed under a powerful pressure, and remain there for a few weeks. As a very cold temperature helps to make the caoutchouc more compact, it is advisable to carry on the work of cutting threads in winter, and to place the presses in the open air.

The plates, when taken from the press, should be entirely smooth and of uniform thickness. They are then brought to the cutting machine. This consists of a shoulder upon which the plate is fixed horizontally by means of sharp points. This shoulder revolves on its axis, and, at the same time, progresses forward. A knife, moving quickly to and fro, cuts a spiral band from the plate, the thickness of which, of course, depends on the speed with which the plate approaches the knife.

A stream of water falls steadily upon the knife to prevent the caoutchouc from sticking to it. The long band obtained by the operation is then cut into square cords. In many factories the neck and bottom of the bottles are cut off and the remainder softened in boiling water, then drawn over a wooden cylinder covered with a thin layer of caoutchouc. This cylinder while revolving is raised a certain distance and moves against a knife, placed vertically, which cuts a spiral band from it. This is a simpler manner of obtaining a band of raw caoutchouc, but the cords cut from it are not quite so solid as those manufactured from the pressed halves of the bottles.

The bands, obtained by either of these methods, are now cut into cords or threads by machines, as they do the work with much more regularity than is possible by hand, and save a great deal of time.

The simplest manner of cutting the bands into threads, is between two steel rollers, in the circumferences of which are grooves as broad as the threads to be cut, and so arranged that the upper roller covers every groove in the lower one. The sheet of caoutchouc passing through between these rollers is cut into a corresponding number of equally wide threads which are wound upon reels. The dividing of the sheet of caoutchouc by this machine must be called *crushing* rather than cutting, and, if the threads are to be cut smooth, it is necessary that the edges of the grooves should be extremely sharp.

More complicated, but more effective, is the thread cutting machine, consisting of a horizontal shaft with circular knives, separated according to the width of the threads to be cut. Above this shaft, holding the knives, is placed a roller with narrow grooves into which the knives penetrate slightly.

A pair of smooth rollers grasp the sheet to be cut into threads and carry it through between the grooved roller and the knives, which revolve with as much velocity as possible and cut the sheet into a corresponding number of threads. They are then passed through between glass rods and wound upon reels.

Manner of Cutting Square Cords from Vulcanized Caoutchouc.

Although threads cut directly from the raw material are the toughest, their use is limited to certain purposes, as they can never be obtained of any considerable length, and have the further disadvantage of not being vulcanized. Prepared caoutchouc, either by itself or compounded with sulphur for vulcanizing, must always be used for long threads or for such as are to be prepared from vulcanized caoutchouc.

The plate from which long vulcanized threads are to be cut is first vulcanized and then cut.

At the present time long threads of ordinary or vulcanized caoutchouc are generally prepared from tubing which is divided by a spiral cut. The tubing to be cut is fastened upon a wooden cylinder which fits exactly into the bore of the tube, and this is fastened to a metal screw which gradually moves forward. A knife moving quickly to and fro cuts a spiral strip from the tube, the width of the strip depending on the height of the screw-thread. Rectangular threads are obtained by using a screw with threads of less height than the thickness of the walls of the tubing.

The machines for cutting threads from tubing have recently been much improved, and threads of any desired thickness can now be prepared by using but one screw.

But in whatever manner the threads may be manufactured, it is of the utmost importance that the greatest care should be used in winding them upon the reels to prevent the freshly cut threads from sticking together.

Round Caoutchouc Threads.

For certain purposes round threads are in great demand. But they can only be manufactured from prepared caoutchouc which has been changed into a plastic dough by treating it with proper solvents, and this dough is then pressed through a metal plate having circular holes.

According to the process by *Aubert* and *Gerard*, the purified caoutchouc is cut into small pieces and brought in contact with bisulphide of carbon, alcohol, fusil oil or wood spirit. These do not dissolve the caoutchouc, but disintegrate its particles, so they can be easily manipulated into a uniform paste or dough.

The following is a very suitable mixture for this purpose:—

	Parts.
Caoutchouc	100
Bisulphide of carbon	100
Alcohol 85 per cent. strong	5

The substances are placed in a hermetically closed metallic vessel, and allowed to stand for from 15 to

18 hours. The mass is then pressed through a wire netting with close meshes which retains the particles not entirely swelled up. The dough, which should be of the consistency of thick paste, is then brought into the moulding apparatus.

This consists of a cylinder in the bottom of which are fitted a number of cone-shaped tubes the bore of which corresponds with the diameter of the threads to be formed. A piston, which should join as closely as possible, is fitted into the cylinder. By pushing this slowly forward the dough is pressed out of the above mentioned tubes.

The threads coming from the tubes first reach an endless band of cotton velvet 4 metres (13.12 feet) long. While they are carried away by this band they lose a considerable part of the bisulphide of carbon by evaporation, and obtain thereby a certain degree of solidity. From the velvet band they pass to a second endless band of fine wire gauze which is kept in a shaking motion, while finely powdered talc falls constantly upon the threads. By the shaking motion of the band, the threads are covered everywhere with the talc powder which prevents them from sticking together.

To entirely evaporate the bisulphide of carbon which may still adhere to them, *Aubert* and *Gerard* use a system of endless linen bands consisting of five bands, each 16 metres (52.4 feet) long. They are arranged one above the other and move in opposite directions so that the threads run to and fro.

About ten minutes are required for them to run over all the endless bands, and during this time they lose sufficient of the bisulphide of carbon that they can be wound upon reels without sticking together.

They are wound in the same manner as the loose cotton bands in cotton mills. A funnel stands over tin boxes which all revolve around their axis at the same velocity. The thread glides through the funnel into the box and is wound up in it to a spiral.

When the cylinder from which the dough is pressed is nearly empty, it is filled up again and in this manner threads of any desired length can be manufactured. If threads of a specific diameter are to be prepared, the fact must be taken into consideration, that the diameter of the threads decreases considerably in drying. A thread pressed through a tube having a diameter of one millimetre (0.039 inch), when dry, will have a diameter of only 0.72 millimetre (0.028 inch).

Only threads having at least the diameter of the one mentioned above can be prepared by pressing. If holes of a smaller diameter than one millimetre (0.039 inch) are used, the dough breaks constantly and the work cannot be carried on without interruption. A peculiar physical behavior of caoutchouc is taken advantage of for preparing still thinner threads.

Namely, if a thread of caoutchouc is stretched lengthwise and simultaneously exposed to a temperature of 115° C. (239° F.) it will retain the length to which it has been stretched even after the tension ceases. If the thread which has been dried in this manner is again drawn out lengthwise, and again heated to 115° C. (239° F.) it remains stretched, and by repeating this operation several times threads of a much smaller diameter can be obtained than is possible by cutting or pressing them.

Preparation of Gutta Percha Threads.

The property possessed by gutta percha of being changed into a very plastic mass when heated, makes it very easy to prepare threads of any desired dimensions from it, and rollers as well as presses may be used for the purpose.

If gutta percha threads are to be prepared by pressing an apparatus is used which resembles very much the one used for manufacturing round caoutchouc threads. But the cylinder into which the gutta percha, after it has been heated to 100° C. (212° F.), is brought must be surrounded by a steam jacket to maintain the same temperature in the cylinder.

So as to prevent all interruption of the labor, and to make it possible to press the gutta percha in the form of coherent threads from the narrow tubes, the greatest care must be observed in filling the cylinder in such a manner that the mass is entirely compact and contains no spaces filled with air, as this would cause the threads to break. The simplest way of avoiding this is to bring the softened gutta percha into a cylinder having the same diameter as the press cylinder. The mass is solidly pressed into this and the cylinder of gutta percha, which has thus been formed, is then transferred to the press cylinder.

The other parts of the apparatus are arranged in the same manner as in the one used for manufacturing caoutchouc threads. A number of endless cloths are also used, over which the threads are carried, so that they may cool off and become solid, but they need not be as long as those used in the manufacture of caoutchouc threads. The best plan for cooling them off quickly is to use a powerful ventilator, throwing out a strong current of cold air. Dusting with powdered talc is superfluous, as gutta percha loses all stickiness when cooled off to a certain degree. It is best to wind the finished threads upon rollers of considerable diameter as, when this is done, it is easier to stretch them straight again in case this becomes necessary.

Compounds of gutta percha and sulphur can also be transformed into threads in this manner. They can be vulcanized by burning, but the greatest care must be observed that the threads are wound upon the rollers in such a way that the separate windings do not touch each other.

For preparing threads from gutta percha by rolling, it must first be transformed into a band somewhat thicker than the diameter of the threads which are to be cut.

The rollers used for this purpose are so constructed that half cylinders are cut into each, and the grooves, formed in this manner, stand so close together that their edges touch each other and form cutting edges.

The rollers are placed over each other in such a way that two grooves fit exactly together and form a circular groove.

The plate to be cut into threads is heated to 100° C. (212° F.) immediately before it is passed through between the rollers, and is carried to them over a plate of polished steel. The threads are then cooled off in the same manner as has been described.

By using suitably cut rollers elliptic as well as polygonal threads can be prepared in this manner. The principal point in arranging the rollers is to see that the grooves in the two rollers fit exactly together, as the threads, if this is not the case, will not acquire the desired form.

XXII.

FABRICATION OF CAOUTCHOUC AND GUTTA PERCHA HOSE OR TUBING.

A.—Caoutchouc Hose.

Caoutchouc and gutta percha hose form a very important article as, on account of their pliability and resistance to chemical agents, they are used in many industries. The demands made in practice in regard to caoutchouc hose are manifold, and frequently it is very difficult to supply them.

Hose or tubing for chemists' use, and gas conductors in rooms, should be as thin and pliable as possible, and, at the same time, perfectly gas tight. Hose for conducting compressed air, indispensable in rock drilling, should be able to withstand the pressure of several atmospheres.

It is absolutely necessary that hose which shall answer all reasonable demands, should not kink when used in short bends. This kinking is very annoying as it obstructs the flow of the fluid or gas contained in the hose until it is again straightened out, and subjects the manufacturer to the charge of furnishing unserviceable goods. This evil occurs only in hose with too thin walls, and can be avoided by maintaining due proportion between the diameter and walls of the hose.

Fabrication of Ordinary Hose.

It is the custom now of making the hose from vulcanite, as being more serviceable than that prepared from caoutchouc, as the latter has the fault of becoming brittle and full of cracks, especially if it is exposed to frequent changes of temperature.

A soft mass, obtained by mechanically treating a compound of caoutchouc and sulphur, is used for the purpose of making hose. The mass is first rolled into plates of a thickness corresponding with that of the walls of the hose which is to be prepared, the interior diameter of the latter being determined by an iron mandrel over which the soft mass is formed.

Mandrels of round smooth wire are generally used

for hose of a small diameter, but it is better to use wooden mandrels for hose of a larger diameter, as, on account of their weight, it is very difficult to handle those of iron. But the wooden mandrels must be perfectly cylindrical, and it is advisable to saturate them with hot linseed oil before they are used.

For manufacturing short tubing the mass is cut into bands somewhat wider than the circumference of the mandrel. These are then placed around it and joined together by a gentle pressure. Finally, the mandrel, with its envelope of caoutchouc, is rolled upon a smooth table to give the hose a perfectly cylindrical form.

It is then wrapped in a linen cloth, which remains around it during the burning process. When this is finished, the mandrel is withdrawn and the linen cloth removed from the now finished hose.

If hose of greater length or larger diameter is to be manufactured, the caoutchouc is generally used in the form of a band, which is laid in spirals around the mandrel in such a manner that the edges slightly overlap. By pressing with the fingers and rolling upon the table, they are joined together to a hose which is treated in the same manner as has been described.

But caoutchouc alone is not sufficient where the hose has to bear a great pressure, and it becomes necessary to strengthen it by inclosures of tissues and spirals. Of course this increases the solidity, but decreases the flexibility of the hose to a great degree.

Hose with Inclosures.

Hose with an inclosure of tissues is prepared by first forming a thin hose from pure caoutchouc mass. A piece of the tissue is placed over this, but this must be wide enough to allow the ends to lap over. The tissue before it is laid upon the hose is brushed over with a solution of caoutchouc, and in placing it in position, great care must be observed to prevent the formation of air bubbles, as on the places where such are present, the caoutchouc and tissue do not form a juncture, and experience has shown that, when the hose is subjected to a high pressure, it bursts first at those defective places.

When the tissue has been applied, it is covered with a second layer of caoutchouc mass, so that it is entirely inclosed and can only be seen on the cross-section of the hose.

For hose with wire inclosures, the latter is used in the form of spirals, which are wound over the hose formed on the mandrel, and then covered in the usual manner with a second layer of caoutchouc.

Small hose can also be manufactured with the machine used for pressing threads. But in this case the cylindrical holes through which the caoutchouc dough is pressed, are replaced by openings in which a mandrel of a suitable size is inserted. The latter is hollow and is connected with a vessel containing water. The small hose as it comes from the cylinder is closed by pressing the ends together, and filled with water as quickly as it is formed. This is absolutely necessary, as the hose would collapse if this precaution were neglected, and the sides stick together. The subsequent treatment of the hose is exactly the same as that of the threads. When it is finished it is opened and the water allowed to run off.

B.—Gutta Percha Hose.

Special machines are used for manufacturing gutta percha hose. In their construction they resemble very much the presses used for manufacturing clay pipes. The solidity of gutta percha, and its property of being hard at an ordinary temperature, makes it especially well adapted for the manufacture of hose. Tubes of gutta percha are now frequently used for pump-barrels, and hose of a smaller diameter, as siphons, and for conducting chlorine, as the latter has no effect upon it whatever.

The power of resistance of gutta percha against bursting is also remarkable. Experiments have proved that a hose of 17 millimetres (0.66 inch) diameter withstood for months a pressure of ten atmospheres without being injured in the least.

The construction of the machines used for manufacturing gutta percha hose is as follows:—

A pipe, which determines the outer diameter of the hose to be manufactured, is fitted in the centre of the bottom of a strong iron cylinder. A round mandre having a diameter equal to the interior diameter of the hose sticks in this pipe.

When the cylinder has been filled with the material, a strongly wrought disk, which acts as a piston, is placed in it and pressed slowly forward, but with great force, by a rack and pinion. The charging of the cylinder with gutta percha, softened by heating, is a labor requiring the greatest care, as on this depends the possibility of obtaining long and faultless hose. Namely, the cylinder must be charged in such a manner that the entire space is filled with gutta percha without the occurrence of air bubbles.

To obtain this object the work is done in the following manner: The gutta percha is divided into small balls about the size of a fist and heated; as soon as the mass has become sufficiently soft the balls are placed in the cylinder, two men being required for this work. While one workman throws the heated balls into the cylinder, the second stamps them with a flat pestle into a homogeneous mass. This labor is continued until the cylinder is filled up so far as to just leave room enough for putting the piston into position.

After completing these operations, steam is admitted into the steam-jacket to heat the contents of the cylinder, which may have cooled off somewhat while putting them into the cylinder, and the heat is maintained until the mass is entirely soft. The actual work is commenced when a small test proves that the gutta percha can be worked without difficulty.

The hose upon leaving the machine has a consistency not much greater than flour dough, and must be cooled off as rapidly as possible to an ordinary temperature in order that the gutta percha may harden.

Most of the gutta percha factories pass the hose, upon its leaving the apparatus, through a narrow box, from 10 to 15 metres (32.8 to 49.2 feet) long supplied with running cold water. Practical results have proved this length to be sufficient to cool off the hose so that it will retain its shape.

If hose is to be manufactured of such a length that one charge of the cylinder does not furnish sufficient material, the action of the piston may be stopped when the material in the cylinder is nearly exhausted, and the latter filled anew. The mass is then heated to the necessary degree and the labor continued. Hose more than 300 metres (984 feet) long has been manufactured in this manner.

Only very recently machines have been constructed which permit the manufacture of hose of any desired length. From the manner in which they work, they may be designated as hose forging machines.

A section of hose of a certain diameter and several metres long is prepared with the previously described apparatus. A solid mandrel of a suitable length is inserted in this section, and both are placed in a press, the lower half of which consists of a semi-cylindrical piece of metal corresponding with the outer diameter of the hose. The upper half of the press supports a piece of the same form as the lower, so that both together form a cylinder having a diameter equal to the outer diameter of the hose. The two halves of the cylinder are hollow and can quickly be heated by super-heated steam or hot air to the temperature required for changing the mass into vulcanite.

When the section of hose has been sufficiently burned, the cylinder is opened, the mandrel drawn out, and a new section of hose pressed out from the first apparatus, which is then treated in the same manner. As the burning process always requires considerable time, the press cylinder may be refilled in the meanwhile, and, it will be easily understood, that hose of any desired length can be manufactured in this manner. Lately this apparatus has also been introduced for manufacturing vulcanized caoutchouc hose.

Although the manufacture of vulcanized hose of any considerable length in the usual manner presents many difficulties, especially on account of the limited size of the burning apparatus, it is nevertheless possible to manufacture long vulcanized hose without using the forging apparatus, but it is tedious work requiring much time.

The section of hose must be made so long that a part of it projects from the burning apparatus, and is not burned. To prevent this part from collapsing while the section in the apparatus is burned, a mandrel must be inserted in it. As soon as the section in the apparatus is finished, another section is joined to it, and the hose pushed forward the length of the burning apparatus. Other sections are added in the same manner until a hose of sufficient length has been obtained. This labor, as may be easily understood, takes much time, and, for this reason, it may be recommended to use the forging apparatus in all cases where hose of large dimensions is to be manufactured.

By slightly modifying the press which is used for the manufacture of hose, it may also be employed for preparing solid articles of gutta percha.

For this purpose the front of the cylinder is provided with a pipe to which is fitted the open end of a metal mould, the hollow part of which corresponds with the form the article is to have. The mould must consist of several pieces, so that it can be taken apart and the contents removed, and, besides, must be provided with an opening through which the air contained in it can escape.

In order to form articles in this manner the mould is heated to from $30°$ to $40°$ C. ($86°$ to $104°$ F.) and then fastened to the pipe, and the softened mass of gutta percha forced by a strong pressure of the piston through the pipe into the mould, this being continued until the mass makes its appearance through the air-hole. The mould is then removed and allowed to stand until the gutta percha contained in it is cooled off and congealed. It is then taken apart and the small cylinder, which has been formed by the mass passing into the air-hole, cut off.

If articles made in this manner are to be burned, it can be done without further preparation by simply putting them in the burning apparatus provided they have an even surface upon which they can be placed, and heating them quickly to the required degree of heat. But if they have not an even surface they must be burned in moulds and, as metal moulds for so many different articles would be rather expensive, plaster of Paris is used for the purpose.

XXIII.

FORMING OR MOULDING OF MASSIVE ARTICLES AND HOLLOW BODIES FROM CAOUTCHOUC AND GUTTA PERCHA.

The process of forming massive or hollow articles from caoutchouc and gutta percha has already been indicated in the foregoing pages, and it only remains for us to explain some special manipulations which are required in certain cases. We must here draw a broad distinction between articles to be prepared from pure caoutchouc and pure gutta percha, and such as are to be manufactured from vulcanized masses.

If complicated articles are to be prepared from raw, that is, non-vulcanized, material, the separate parts of them are formed by themselves and are joined together to a whole by a solution of caoutchouc, if this material has been used; or, in case of gutta percha, by rubbing the edges to be joined with a hot iron.

Among the many articles manufactured from rubber, the production of dolls and toys has become an important branch of the rubber industry, as, on account of their indestructibility and softness, they are especially adapted for children's toys. These articles have of late been improved so much that, as far as beauty of form is concerned, they may be called small works of art, and, in fact, such figures are at present frequently used for ornaments in rooms.

Small articles of this kind—for instance, human figures—are pressed from vulcanized caoutchouc in forms of metal, so that the figure is obtained in two halves each a few millimetres thick. These are joined together by a solution of caoutchouc, so that they form a hollow body, and are then burned. But the air inclosed in the figure would expand it so much during the burning process that it would burst. To prevent this a small hole is made in some part of it through which the air can escape, and this, after the burning, is closed by a small cork of caoutchouc dough.

Such figures are also prepared from sheets of vulcanized caoutchouc-mass and by using moulds of type-metal. The sheets are cut with a pair of scissors into suitable shapes and lightly pressed into the mould; this is then closed so that both sheets join together. But before the mould is pressed together tight, a few drops of water are poured into the interior of the article. When the mould prepared in this manner is exposed to the heat of the burning apparatus, the water inclosed in the caoutchouc-mass is changed into steam which forces the sheets apart, so that all cavities of the mould are filled up. When the articles are taken from the mould, which must be done while they are still hot, a small hole is made in them to allow the air to pass into the interior and to prevent them from collapsing while they are cooling off.

The elastic balls used for playing ball are made in a similar manner. First segments of a sphere are formed from a plate of caoutchouc-mass; these are joined together to a ball in a mould of plaster of Paris and burnt while in the latter. The balls are filled with compressed air by a special apparatus to give them the necessary elasticity. This apparatus consists of a small condensing air-pump provided with a manometer. The pipe through which the condensed air escapes ends in a fine needle which is put into the interior of the ball. The air in the ball is compressed until the manometer indicates a pressure of three atmospheres; the needle is then withdrawn and the hole quickly closed with a small quantity of vulcanite-mass, which is vulcanized by bringing it into close proximity with a hot iron.

Hollow articles of any desired form can also be prepared by using the dough of caoutchouc, the preparation of which has been described in a previous chapter. The process is nearly the same as that used for forming hollow articles from plaster of Paris. The moulds used for this purpose may be made of metal, wood, or plaster of Paris, but, when moulds of the last two materials are employed, they must be coated with linseed-oil varnish before they are used, and this must be repeated until the mould absorbs no more varnish.

Moulds consisting of more pieces than one are put together in the same manner as those used for moulding articles from plaster of Paris, and the dough of caoutchouc is then poured into the hollow part. In doing this the mould is swung to and fro in such a manner that the interior wall is covered with the thick fluid mass, and then the excess of dough is allowed to run off. But, with some experience, the latter is not necessary, as a skilled workman can estimate the exact quantity of dough required for a mould of a certain size.

To accelerate the evaporation of the solvent from the dough, it is advisable to fit a pipe into the form through which air is blown into the interior. By this the vapors of the solvent are quickly carried off. Finally, the form is placed in a warm room for drying out.

A suitable quantity of finely powdered sulphur or other vulcanizing agent may be mixed with the dough before it is formed into articles, and, if this has been done, it is only necessary to place the articles which remain in the mould in the burning apparatus and to burn them. If pure caoutchouc dough has been used, they can be vulcanized by a simple process.

For this purpose they are dusted with finely powdered sulphur, and some of it is also blown into the interior of them. They are then burned in the usual manner, or they are vulcanized by using chloride of sulphur.

Small solid balls of vulcanized caoutchouc are required for sewing machines. These are prepared from vulcanite mass by stamping and burning them. A peculiar process is used in many factories for manufacturing them from pure caoutchouc. A block of it is pressed against a grater, which revolves as quickly as possible, and by this is cut into exceedingly fine shavings which can easily be balled together.

Balls are formed from these shavings with the hand. They are tightly pressed in metal moulds, then brought into a somewhat smaller mould and in this subjected in the cold to as strong a pressure as possible. They are very compact, and must be heated to 40° C. (104° F.) to restore to them their elasticity. Balls manufactured in this manner are especially adapted for bearers under powerful stamping presses, as they drive the stamp back with great force.

The small balloons, so much used as toys, are also specialties of the rubber industry. Very small balloons are made from a clear solution of caoutchouc. A glass balloon serves as a form; a certain quantity is poured into this and distributed over its entire inner surface by swinging it to and fro, and the excess of solution is then allowed to run out. When nothing more drops out of the glass balloon, this is again placed so that the neck is uppermost and air is blown into the interior of the balloon to accelerate the evaporation of the solvent.

To detach the balloon from the glass, to which it adheres quite tenaciously, the lower edge of it is carefully loosened from the glass and then air blown in between the film and the glass. The thin film, of which the balloon consists, becomes entirely detached from the glass and can be removed in the form of a bag.

The caoutchouc plates prepared upon glass plates, as described in a former chapter, may be used for somewhat larger balloons, such as are employed at meteorological stations for ascertaining the direction of the wind. As most of these balloons have a volume of a few litres only, they are generally filled with illuminating gas and then closed.

Coating of Wires with Gutta Percha.

Of all industrial applications for which gutta percha is used, there is none of greater importance than the coating of wires with it, as on this depends the establishment of submarine telegraph lines. There is no other body, suitable for this purpose, which is such a non-conductor of electricity. Many comparative experiments have proved that a wire becomes insulated by simply allowing a coat of an ordinary solution of gutta-percha to dry upon it.

As such a coat not only insulates the wire, but is also an absolute protection against rust, it may certainly be recommended to provide wires, which are to be used for telegraph lines inside of houses, with a very thin coat of such a solution. Such wire would almost last forever.

As a coating of gutta percha not only secures complete insulation, but also protects the wire against the action of sea-water, it may be said without exaggeration, that, if the properties of gutta percha had not been known, it would have been impossible to lay submarine telegraph lines.

Insulation of telegraph wires by coatings of gutta percha demands attention to two very important conditions: The wire must lay exactly in the centre of the cable or cylinder, and the covering be perfectly continuous. The smallest crack would allow the entrance of sea-water and eventually destroy the insulation.

As the manufacture of telegraph cables requires large factories, we will not discuss the subject very fully, but confine ourselves to describing the process of coating wire with gutta percha, large quantities of which are used in the manufacture of electric and magnetic machines.

The apparatus used for the purpose consists of a cylinder containing the gutta percha which has been softened by heat and is pressed forward by a piston.

The soft mass passes out of an opening which determines the thickness of the gutta percha cylinder. Below this opening is a die of metal, with a hole just wide enough to allow the wire to be coated to pass through without much friction.

When the piston, which acts upon the softened material presses forward, a cylinder of gutta percha is

FORMING HOLLOW BODIES. 221

pressed out of the respective opening, and this, on account of the strong friction, carries the wire with it and incloses it entirely.

As the material must be heated so far that it can be pressed out of the small opening without the use of too much power, provision must also be made for sufficiently cooling the mass off before it is wound upon a drum. All that is necessary in this case, is to carry the wire through a channel several metres long, which is kept constantly filled with water, and then to wind it upon a large drum in such a manner that the separate windings lay along side each other.

The accompanying illustration (Fig. 6) shows such an apparatus of the most simple construction. C is a

Fig. 6.

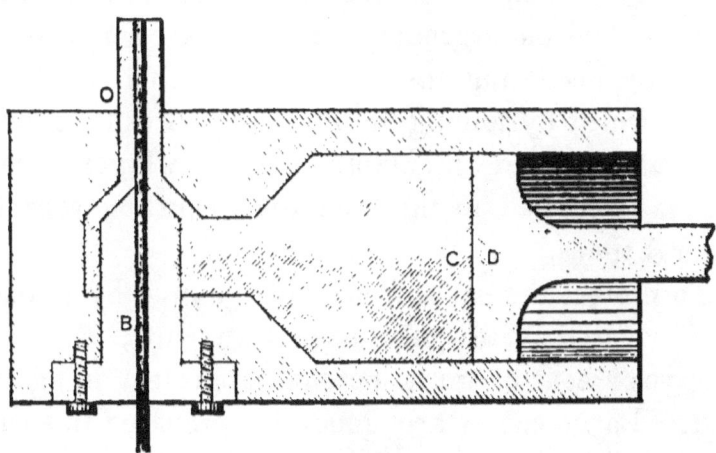

cylinder containing the softened gutta percha; D is the piston by which it is pressed forward; B is a metal cylinder in the bore of which the wire sits. This is

19*

exactly opposite to the opening, O, through which the gutta percha is pressed. The wire which is to be covered with it, is wound loosely upon a drum, and is drawn forward in consequence of the high pressure which the compressed material exerts upon it.

If several wires coated with it are to be formed into one cable, several such dies of metal in which the wires sit are fastened to the cylinder from which the gutta percha is pressed. The wires which are to be formed into one cable are passed through a cylinder while they are still warm, so that their coatings adhere to each other. The diameter of the cylinder should be such that the wires, in passing through it, are gently pressed against each other.

But a simple coating with gutta percha does not suffice for cables, which are to be immersed in water or to be laid underground. In such a case it serves only for insulating the wire. The cable which has been formed by joining the insulated wires together is generally covered with Manilla hemp. A layer of gutta percha is applied to this and the operation repeated several times.

To protect the cable from being gnawed by animals, it is covered with galvanized iron wire; this also receives a coat of gutta percha to protect it against rust. Large cables are generally prepared in such a manner that six copper wires coated with gutta percha lie around a centre wire, which is also coated, the seven wires together with their coatings forming a cylinder having a diameter of from 8 to 10 millimetres (0.31 to 0.39 inch).

But long submarine cables must be still further secured against breaking, and for this purpose are covered with suitable materials (Manilla hemp, galvanized iron wire, etc.), until they have a diameter of 30, aye, even 40 millimetres (1.18 to 1.57 inches).

To give an idea of the dimensions of some of the submarine cables in use, we will only mention that the cable which was laid in 1866 between Europe and North America weighs nearly 31 cwt. per knot, and is altogether about 4000 knots long. The longest submarine cables which have been laid thus far are: The one between *Ireland* and *Newfoundland* 1896 English miles long, and that from *Valencia* to the same place 1900 miles long. The cable between *St. Vincent* and *Pernambuco* measures 1953 miles, and the one between *Brest* and *St. Petersburg* as much as 2584 miles.

XXIV.

FABRICATION OF CAOUTCHOUC SPONGES.

The caoutchouc sponges belong to those specialties of the rubber industry which, without doubt, have a great future before them. These products, which were first brought into commerce under the above name by English factories, have exactly the appearance of porous bathing sponges but, as far as softness and durability are concerned, surpass the latter by far.

The English manufacturers have thus far succeeded in keeping secret their process of manufacturing these sponges, but it is very likely the same as is very successfully used at the present time by French and German manufacturers. At least, the sponges produced by the latter are equal in appearance and quality to the best English products.

According to experiments we have made on this subject, the caoutchouc sponges can very well be manufactured in the following manner: A thickly fluid solution of caoutchouc in benzol, chloroform or bisulphide of carbon is placed in a high vessel of tin of a prismatic form so that it stands a few centimetres high in it, and is then heated to above the boiling point of the respective solvent.

The mass becomes more tenacious and thickly fluid, in consequence of the evaporation of the solvent, it becoming more and more difficult for the steam-bubbles to break their way through it, and the effect of this is that it remains behind, porous and full of holes. When caoutchouc dough is employed, and the precaution used of heating it very slowly, sponges are obtained which have very fine pores and in regard to softness by far surpass the finest bathing sponges. The finished product is then vulcanized by plunging it in a solution of chloride of sulphur, and may be further provided with a suitable base of hard rubber so as to give it a handy shape for use.

The disagreeable odor of these sponges, which is perhaps the more perceptible on account of the porous nature of the mass, than is the case with other vulcanized articles, is the only thing which prevents them from being introduced into general use, and this odor is doubly disagreeable, when we consider the purpose for which they are to be used.

It must be, therefore, to the interest of all manufacturers, to remove this odor as much as possible. We have found animal charcoal to be the most suitable agent for this. The sponges are simply wrapped up in tissue-paper, and placed in a vessel filled with powdered animal charcoal. In a few weeks—especially if the vessel is put in a warm place—the sponges will have lost nearly all odor, and the last traces of it may be removed by thoroughly washing them.

XXV.

FABRICATION OF RUBBER SHOES.

In the course of time this article has passed through a peculiar process of development. The first so-called gum shoes consisted of a single piece of caoutchouc, and were manufactured in the same manner as the bottles of caoutchouc. Clay forms, having the shape of a last, were coated with the milky juice of the caoutchouc tree and dried over a fire. The congealed coating was then drawn from the last and formed the shoe.

The shoes prepared in this manner were very durable, but ugly, and at the same time possessed the disadvantage of tightly inclosing the feet, and in a short time producing in them the feeling of unbearable heat.

Rubber shoes are now prepared in the following manner: A tissue is coated with a sufficient quantity of caoutchouc to prevent the water from penetrating, and the shoe receives such a form that it does not hermetically inclose the leather shoe over which it is to be drawn. We will here mention that under the name of rubber shoes products are brought into the market, which do not deserve the name, as no caoutchouc whatever, but only elastic varnishes colored black, are used in manufacturing them.

The genuine rubber shoes, as found in commerce at the present time, are prepared by coating a quite loose tissue with a very thin layer of vulcanite mass, which has been colored with lampblack. The separate parts, which are to form the shoe, are cut from this prepared substance and pasted together with caoutchouc solution over an iron last. The sole is cut from a somewhat thicker plate.

The shoes are left on the last over which they have been pasted and subjected to burning; but first they are provided with a coat of asphaltum varnish to give them a fine lustre. As it is also impossible to manufacture these shoes so that they will not make the foot hot, all kinds of contrivances have been employed to create ventilation. For instance, the upper part of the sole has been provided with small holes, or it has been composed of scale-like pieces; but, of course, by this the shoe loses more or less its property of preventing the water from penetrating.

XXVI.

FABRICATION OF WATERPROOF TISSUES WITH CAOUTCHOUC.

BEFORE caoutchouc and gutta percha were known it was very difficult to prepare waterproof tissues, and, as is well known, they were exclusively manufactured by applying a coat of varnish to closely-woven tissues (oil-cloth). Although this material answered quite well the purpose for which it was intended, it had one great disadvantage, namely, the tissue lost a great deal of its pliability, and in the course of time the coating became so brittle that the material commenced to crack, and in a short time became entirely useless.

As soon as caoutchouc and its solubility became known, it was at once used for making tissues waterproof. It is claimed that the Englishman, *Mackintosh*, first manufactured such material, and it was therefore called by the name of the inventor; but at the present time the term "*waterproof*" is generally used in commerce for each article, this, perhaps, being done for the reason that very little caoutchouc is used in many compositions for the manufacture of waterproof materials.

Although the tissues manufactured by *Mackintosh* possessed the advantage of being very durable, they had also the disadvantage of being very weighty, thick, and expensive. *Mackintosh* prepared his waterproof goods by placing a thin sheet of caoutchouc between two tissues, and passing the whole through between heated rollers.

By this manipulation the caoutchouc was so strongly heated that it became soft, and was pressed into the meshes of the fabrics, cementing them firmly together. Many experiments were made to improve *Mackintosh's* process by decreasing the weight and thickness of the materials, but this has only been accomplished since the introduction of the new process of working caoutchouc, which makes it possible to prepare very thin plates from it.

Dumas already proposed to prepare very thin plates by allowing a solution of caoutchouc in ether to run down over heated rollers. The ether would evaporate, and the thin, soft plate, which could easily be detached from the polished roller, was to be spread upon the tissue, this to be covered with the second tissue and both joined together by rolling.

It is absolutely necessary that the coating of caoutchouc should be as thin as possible, if it is desired to prepare an article which will answer all reasonable demands. So much progress has been made in cutting cylinders of caoutchouc into thin plates, that they can be furnished not much thicker than a sheet of writing paper. A successful attempt has been made to coat only one side of a tissue with such thin plates, but this unfortunately has the disadvantage of making the material heavy and of being rather expensive. A decided improvement in the manufacture of waterproof materials has only been made since the process of obtaining caoutchouc in a very soft form by mechanical treatment became known, but the greatest perfection in the manufacture of this important article has only been reached since the introduction of vulcanite.

The first improvement of *Mackintosh's* method consisted in using but one layer of tissue and, by making the sheet of caoutchouc as thin as possible, the volume of the material was considerably decreased. The manner of manufacturing these water-proof fabrics was a very simple one and was carried on as follows:—

The crude caoutchouc was first brought into as homogeneous a mass as possible by passing it through between rollers and finally rolled into very thin sheets. The mass, when it comes from the rollers, possesses considerable stickiness, and this was made use of for joining the tissue with the caoutchouc. This was done by passing the tissue together with the caoutchouc through between the rollers when the latter was rolled for the last time.

Before we proceed to discuss the more modern methods of preparing water-proof fabrics, we will say a few words about the character of the tissues to be used. The material may be either silk, wool or cotton, and as the tissue is simply the bearer of the substance which makes it water-proof, either loosely or closely woven fabrics may be used. In the first case a considerable quantity of caoutchouc is required to make it waterproof and only a small one in the latter case.

Most manufacturers have found that it is best to use closely woven fabrics for the manufacture of water-proof tissues, as these require less caoutchouc and besides have the advantage of being firmer. Therefore strong cotton goods are generally used at the present time for manufacturing finer articles, such as water-proof coats and cloaks. The principal requisite of such materials is that, they should be as even and smooth as possible, as even the smallest knots in the tissue exert an injurious effect upon the quality of the goods to be manufactured.

At the present time, as has been mentioned, very thin coatings of caoutchouc can be prepared. If there should be knots in the tissue they of course would be covered with caoutchouc and the fabric would seem to be of excellent quality as long as it is not used. But, if a garment made of such material, is used for a short time, the caoutchouc will commence to peal off wherever the knots are.

Fabrics of cloth and vulcanized caoutchouc can be prepared by rolling the soft vulcanite mass (consisting of caoutchouc and sulphur) into very thin sheets, and by passing them together with the cloth through between heated rollers. The object of hot rollers is twofold, namely to force the caoutchouc mass firmly into the meshes of the tissue and, if the proper temperature is used, also to burn the vulcanite mass.

One pair of rollers, heated to the temperature required for burning, may suffice for the work, but very careful and skilled workmen are then required. Experience has shown that it is much better to use two pairs of rollers, the first of which is heated at the utmost to $120°$ C. ($248°$ F.), while the second is heated to the temperature required for burning the caoutchouc mass. Finally, the finished fabric is rolled upon rollers as soon as it has become sufficiently cooled off, and should be used as soon as possible.

It is also very difficult to obtain uniform products with this apparatus, and experiments have, therefore, been made to use caoutchouc which has been vulcanized. But as this, as is well known, cannot be stuck together, and will not combine with the tissues by pressing, it is necessary to have recourse to certain manipulations to obtain the desired object.

According to the method recommended by *Johnson*, very thin plates of vulcanized caoutchouc are prepared, and these partly desulphurized by boiling them for some time in caustic soda. After they have been boiled they are first washed in water containing some hydrochloric acid (to remove the last traces of the alkali), next in pure water and then dried.

The plates, thus prepared, are roughened by passing them over a roller covered with emery paper revolving with great velocity (800 to 900 times a minute). The object of this is to facilitate the union of the caoutchouc with the tissue.

The plate is then coated with a solution of caoutchouc and placed upon the tissue and both together are passed through between the rollers by which the union of the caoutchouc with the rubber is effected.

The articles manufactured in this manner are of an excellent quality, but, on account of the process being so very complicated, rather expensive. Since the process of completely dissolving caoutchouc has become known, solutions, or a dough of it, are almost exclusively used for the manufacture of water-proof tissues, and generally they are only coated on one side.

Although the labor seems to be a very simple one when solutions are used, nevertheless many difficulties present themselves in the execution of the operation. Frequently the very volatile oils, which are obtained by distilling coal tar—light coal-tar oil or coal-tar benzole—are used as solvents on account of their cheapness. It is true that these solvents volatilize very quickly, but they are always mixed with certain, although small quantities, of less volatile products, which remain behind after the volatilization of the lighter oils, and the odor of which adheres to the caoutchouc so that it is perceptible for years. This odor—agreeable to no one—is so repulsive to some persons that they will not use a garment manufactured from such tissue.

A further evil of using pure solutions is that the layer of caoutchouc, which is left behind, remains sticky for some time, and, for this reason, a garment, prepared from it, cannot be folded up, as the surfaces would stick together so tight that the folds could not be separated.

All these evils are now removed by using a dough of caoutchouc which also contains the substance required for vulcanization, and by subjecting the freshly coated tissue to the burning process.

Such solutions, or more correctly masses, of dough used for this purpose are at once compounded with the quantity of sulphur required for vulcanization. The simplest manner of doing this is to saturate the bisulphide of carbon, which is to be used for dissolving the caoutchouc with sulphur, or by working very finely powdered sulphur into the mass. The principal requisite for a correct execution of the work is, that the mass should be entirely homogeneous, a property which can be imparted to it by a suitable mechanical treatment, as has been explained in a previous chapter.

Spreading the Caoutchouc Mass.

An especially constructed apparatus is required for spreading the caoutchouc mass upon the tissue in as uniform a manner as possible, and to join both intimately together. It makes very little difference which of the various apparatuses recommended is used, as they all answer the purpose for which they are intended. A thoroughly experienced workman is the principal requisite for the work.

Fig. 7 represents an apparatus well adapted for all purposes.

Two rollers of a small diameter—generally 18 or 20 centimetres (7.08 or 7.8 inches)—are placed in

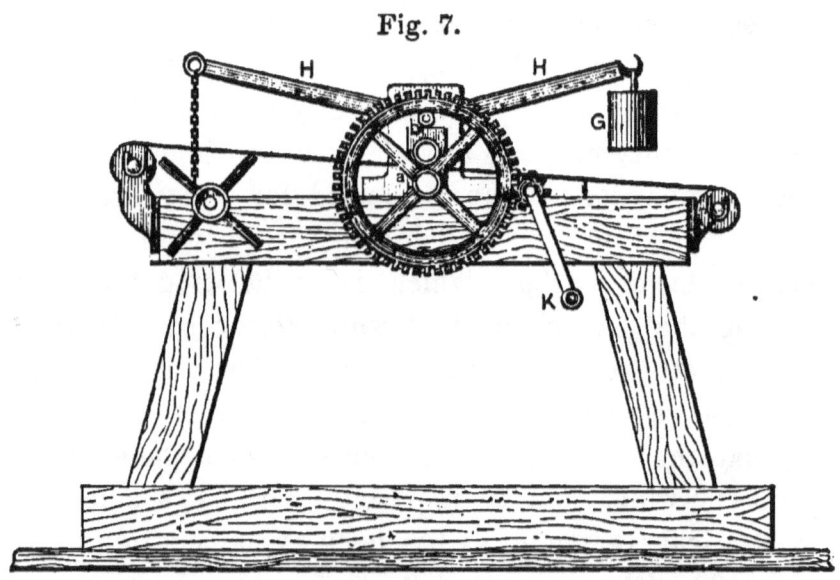

Fig. 7.

brasses resting upon a suitable frame. The lower roller is made to revolve, and is set in motion by cogwheels and the crank K, with which they are connected. The upper roller has square arbors which fit into the brasses, and therefore does not revolve.

The object of the upper roller is to regulate the thickness of the solution which is to be spread upon the tissue, and for this purpose is provided with a peculiar mechanism. Two levers, H, H, having their fulcrum at b, and loaded with weights G, press with a certain force the upper roller against the lower one, this force increasing in power the more horizontal the position of the levers becomes.

As may be seen in the accompanying illustration the levers are double armed, and on the other end are connected with chains, which can be wound upon pulleys. The less the pressure shall be, which is to be exerted upon the upper roller, that is to say, the thicker the caoutchouc mass is to be spread upon the tissue, the tighter the chains are drawn, and the more the arm of the lever, provided with weights, is raised.

The tissues to be transformed into water proof fabrics are wrapped around a roller, and are wound upon a cylinder after they have been provided with the solution, and sufficiently dried, so that the separate folds will not stick together.

We will here make a few remarks in regard to the tissue. The closer woven it is, the thinner the caoutchouc can be spread upon it, as in this case thin solutions can be used without fearing that they will soak through, which must by all means be avoided. For looser tissues a thicker solution, or dough, has to be used, as thin solutions would soak through them, and a properly prepared water-proof fabric should show the coating on one side only.

The thinner the solutions are the more beautiful and uniform will be the coating, and for this reason it also becomes necessary to spread the solution several times upon the tissue to make it entirely water-proof. But for the first coat a solution must be used so thickly fluid that it will not soak through. In many factories the solution, or the dough, is still applied by means of ladles, which the workmen fill from a vessel containing the material.

As the tissue advances between the rollers, he pours the solution upon it, and it may be easily seen that it requires considerable skill to pour the solution upon the tissue in such a manner as to spread exactly the right quantity upon it. And, as besides the solutions always contain fluids, the vapors of which exert an injurious effect upon the health of the workmen, the manufacture of waterproof tissues in this primitive manner deserves to be called very unhealthy work.

To protect the workmen from the evil effects of the vapors, the tissue, after it has been coated, is brought into a room, from which the vapors are constantly pumped.

But this gives only temporary protection, as the workmen who lift the fluid from the vessel and spread it upon the tissue to be sure do not suffer from the vapors which are formed *behind* the rollers, but are exposed to those developed on *the other side* of the apparatus.

A very simple contrivance, represented by Fig. 8, may be used for removing this evil and at the same time for spreading the mass as uniformly as possible upon the tissue. This apparatus consists of a tin-box K, having the shape of a three-sided prism, and serves

FABRICATION OF WATERPROOF TISSUES. 237

for the reception of the solution or dough. The box is placed immediately behind the rollers W. The lid of the box catches into a gutter filled with water which prevents the escape of vapors into the workroom.

A cock H is fitted into the lid, which is only opened at the commencement of the labor, and allows air to enter into the interior of the box as soon as the contents commence to run off. On the lower edge of the box is a slide which closes a slit in the box through which the fluid escapes.

Fig. 8.

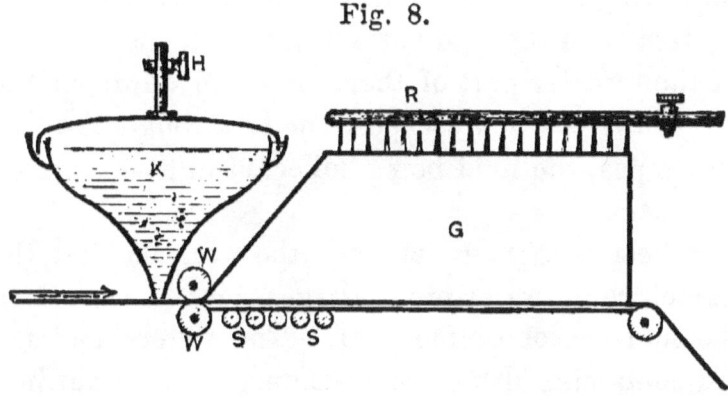

Close to the rollers stands a tin case G, having the form of a house with a steep roof. It is somewhat wider than the tissue which slides along upon the bottom of it, and passes out through as narrow an opening as possible. On top of the roof is a pipe R, provided all around with fine holes and connected with a reservoir constantly filled with cold water. On the sides of the box are several pipes for carrying off the fluid dripping down.

The apparatus works as follows :—

A stream of the solution as wide as the tissue to be coated flows through the slit in the vessel which has been opened just as wide as is necessary, the rollers spreading it upon the tissue in an entirely uniform layer. The part of the bottom of the box nearest to the rollers, into which the tissue passes, is heated by hot water contained in the pipes S. Evaporation of the solvent commences at once and the vapors rise up to the roof. But this is constantly cooled off by the water falling down like rain, and the vapors, on coming in contact with the cold surface, are condensed, or, at least the greater part of them, and run down on the inside of the box and escape from it through the discharge-pipes, the fluid being collected in flasks placed under them.

The best plan is to arrange the work so that the manufacture of water-proof tissues can be carried on in the cold season of the year. The rollers and the vessel containing the solution should then be put in a heated room, while the tin-box, in which the solvent is again condensed to a fluid, is placed in the open air. With such an arrangement the workmen do not suffer from the injurious vapors, the work can be carried on without interruption and the greater part of the solvent may be regained.

As a certain quantity of the solvent still adheres to fabrics which have recently been finished, and causes them to be sticky, it is advisable not to wrap them up immediately after they come from the apparatus, but to keep them stretched out for several days as smoothly as possible.

If tissues for the manufacture of garments are to be prepared, a certain quantity of lampblack is added to the caoutchouc-mass before its mechanical treatment is commenced, and worked with it for the purpose of imparting a uniformly black color to it.

A coating of light brown color is obtained by using caoutchouc in a pure condition. By mixing it with sulphur and heating the ready tissue to the temperature required for vulcanization, the coating will show the peculiar gray coloring of vulcanized caoutchouc.

Manufacture of Tissues with Caoutchouc Inclosures.

Cut plates of caoutchouc are not now used for manufacturing water-proof tissues with caoutchouc inclosures, such as resemble those first manufactured by *Mackintosh*, solutions being generally used for the purpose. The tissues are united with the assistance of an apparatus represented by Fig. 9.

The two tissues having been wound upon the rollers W, are unrolled in the direction indicated by the arrows and pass through between two rollers placed in a horizontal position. A vessel resembling the one represented in Fig. 8, contains the solution of caoutchouc, and is placed exactly over the rollers. The solution is allowed to run from this by opening the slide in the bottom of it as wide as may be necessary, and on reaching the tissues, it is forced into their meshes by the rollers. A second pair of rollers,

Fig. 9.

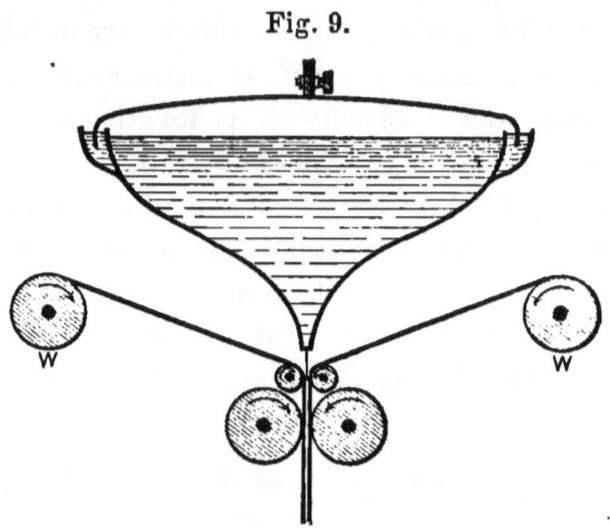

placed beneath the others and heated by steam, effect the evaporation of the solvent. The finished tissues are hung up for a few days to allow them to become completely dry.

Deodorization of Water-proof Fabrics.

It is a very disagreeable fact that the odor of the solvent adheres for a long time to tissues made waterproof by means of pure caoutchouc. It, therefore, becomes necessary to remove this odor as much as possible. It is not sufficient simply to expose the tissues to a higher temperature. More satisfactory results may be obtained by employing a change of air simultaneously with a high temperature, for instance, by conducting a current of hot air through the room in which the tissues are hung up.

It is a well-known fact that certain bodies which volatilize with great difficulty when left to themselves, do so very readily when brought in contact with hot steam, and this process can also be used for removing the disagreeable odor from tissues coated with pure caoutchouc.

The simplest plan is to use the saturated steam as furnished by the steam boiler of the factory. The tissues are suspended in a properly arranged room, and the steam is passed into the latter through several openings. A rather narrow escape pipe for the steam is placed at the other end of the room, and provision must also be made for the escape of the condensed water. A pressure of but little more than one atmosphere is sufficient for all purposes, and the tissues will be entirely deodorized after they have been exposed for some time to the action of the steam. Oil of turpentine is frequently used as a solvent of caoutchouc, when goods of a less fine quality are to be manufactured. In this case some odor will always adhere to the tissues, as some small quantities of empyreumatic substances having a very disagreeable odor are generally mixed with it, and it is a very difficult matter to remove this odor even by treating the tissues with steam.

The same holds good in regard to the coal-tar oils, and for this reason the respective fluids should always be tested before they are used, and those showing the odor in a remarkable degree should either not be used at all or subjected to a second rectification.

XXVII.

MANUFACTURE OF WATER-PROOF TISSUES BY MEANS OF CAOUTCHOUC COMPOSITIONS.

The use of pure caoutchouc, or vulcanite mass, for the manufacture of water-proof tissues is always a rather expensive business, as, besides the considerable labor required, a large part of the solvent is lost even when all imaginable precautions are used.

To enable the manufacturer to prepare water-proof tissues at a lower cost, it has been tried to partly use less expensive bodies than caoutchouc, and tissues are now manufactured which actually contain no caoutchouc whatever. It is scarcely necessary to say that in regard to quality such fabrics cannot be compared with those manufactured from pure caoutchouc.

Coal tar and boiled linseed-oil have proved themselves to be good substitutes for a part of the caoutchouc, as they furnish compositions very well adapted for many purposes, for instance, for the manufacture of so-called rubber shoes; they being now more frequently used for this purpose than pure caoutchouc.

To prepare masses which, besides caoutchouc, are to contain linseed-oil, the latter is heated until decomposition takes place. For this purpose it is placed in a boiler which must be large enough to contain at least three times the quantity of oil, as it expands very much during the heating.

It should be heated as quickly as possible to 150° or 160° C. (302° or 320° F.) and kept at that temperature for several hours. The fire is then sufficiently increased to give the appearance of boiling to the oil which coincides with its decomposition. The heating is continued until a sample of it, taken from the boiler by a wooden spatula, runs off in long, viscid threads.

The boiled oil now possesses the properties of a quickly drying varnish, and until it is used must be protected against the action of the air, to prevent it from drying in. For this purpose it is put in a vessel, after it has become cold, and is covered with a layer of water.

The purified caoutchouc is dissolved in oil of turpentine, which is generally used as a solvent in this case, and the solution compounded with a certain quantity of boiled oil, the latter depending entirely on the pleasure of the manufacturer, as the solution can be mixed with any desired quantity of it. When the first coating is dry, a second or a third is applied as may be found necessary, and finally one consisting of boiled oil alone, to which has been added some lampblack or any other coloring matter.

The tissues can be coated by using the same apparatus employed for preparing fabrics with pure caoutchouc, or, by stretching them over a frame and applying the composition with a flat brush.

When such tissues have been prepared with proper care they are especially well adapted for the manufacture of rubber shoes. The separate parts of the shoes are cut from it according to patterns, and pasted together over a last with a solution of caoutchouc. The soles are manufactured either of vulcanized caoutchouc or may be made of the same tissue, but in the latter case either very thick tissues are used or several layers of the fabric, employed for the uppers, are joined together by pressing them while they are still sticky, in order to make the soles more durable.

If it is desired that the boiled oil should dry as quickly as possible, a small quantity of sugar of lead is added to it before it is boiled; one per cent. of the weight of the oil being sufficient for the purpose.

When coal tar is to be added to the caoutchouc, it must first be boiled, until a mass of the consistency of Burgundy pitch is formed. While still hot, it is kneaded together with and worked in the same manner as pure caoutchouc. This composition can be vulcanized by adding sulphur to it during the mechanical treatment, and subjecting it to the burning process, but the quantity of sulphur added must be somewhat larger than that required for pure caoutchouc.

Water-proof tissues may also be prepared by using a caoutchouc lacquer, as has been described in a previous chapter. In regard to pliability and lustre, the quality of the coating will largely depend on that of the varnish used.

XXVIII.

FABRICATION OF ELASTIC WEBBINGS.

What is now understood by elastic webbing can only be manufactured by using caoutchouc, and these fabrics have become of great importance, as they are indispensable in the manufacture of shoes. Plates or threads of caoutchouc may be used for manufacturing them, but the first are of only secondary importance, as threads are employed for preparing the larger part of these fabrics.

Closely woven tissues, the separate threads of which possess a considerable degree of elasticity, must be used for manufacturing them with the assistance of caoutchouc plates. They are stretched over a frame, brushed over with a solution of caoutchouc, and then covered with a plate of it, which has been stretched in the same manner as that used for threads described in a former chapter. Upon this is placed the second tissue, which has also been first brushed over with a solution of caoutchouc. After the material has become dry, it is exposed to a temperature of from 60° to 70° C. (140° to 158° F.) in order to restore the elasticity of the caoutchouc. If plates of vulcanized caoutchouc are to be used, they must first be partly desulpherized and roughened with pumice stone to make it possible for the solution to adhere to them. They are then worked in the same manner as plates of the ordinary material.

The threads to be used for webbing must first be subjected to stretching. For this purpose they are placed in warm water and allowed to remain there for some time, whereby they acquire a great degree of ductility. They are then wound upon reels, being strongly stretched during the process, and exposed to as cold a temperature as possible, until they have lost all elasticity. For this reason it is advisable to arrange the work so that the operation of stretching the threads can be done in the cold season of the year.

Threads, which have been properly stretched and cooled off, should not contract when taken from the reels nor show any perceptible elasticity. Should a thread break during stretching, the apparatus is stopped, the place of rupture cut obliquely with a pair of sharp scissors, and the freshly cut surfaces joined by pressing them together.

The weaving of caoutchouc threads is done by forming a kind of net consisting of six or seven threads of any kind of yarn around each thread, and these overspun threads are then joined together by a woof. They may also be wound upon the beam as a chain and provided with a woof of ordinary yarn.

Another method of preparing elastic webbings, but furnishing a less durable article, consists in placing the threads parallel along-side of each other upon a tissue which has been brushed over with a solution of caoutchouc, and covering them with a similar tissue. They are then joined together by passing them through powerful rollers.

Tissues prepared in either manner are finally finished by passing them through rollers heated by steam, which impart to them a temperature of from 60° to 70° C. (140° to 158° F.). The caoutchouc, which has become inelastic by having been exposed to a low temperature, assumes by this its original elasticity. The stretched threads contract and effect also the contraction of the tissue joined with them.

When threads of vulcanized caoutchouc are to be used in the manufacture of elastic webbing, the process must be changed in a corresponding manner, as vulcanized caoutchouc does not possess the property of retaining the length given to it by stretching, but will contract to its original length when the tension is discontinued.

In this case the loom must be so arranged that the threads are firmly stretched while they are woven and the finished woven material must also be subjected to such tension. The power acting upon the tissue is only released when the latter is entirely finished, when it will contract to the original length of the threads.

Manufacture of Caoutchouc Stamps.

A vulcanizing apparatus with lamp and thermometer, such as dentists use, is required for this, and besides an iron chase in which the types are confined. The types are oiled in the usual manner and the mass is then poured over them. The stereotype plate is not allowed to become dry, but is laid upon a plate of vulcanized caoutchouc. Both plates are then pressed between two iron plates, the caoutchouc being pressed in this manner into the stereotype plate. A few sheets of paper are laid between the caoutchouc and the iron plate to prevent the first from sticking to the latter. The whole is then brought into the water of the vulcanizing apparatus, and, after the cover has been screwed down, heated to 152° C. (305.6° F.). When it has become cool, the form is taken from the apparatus and the caoutchouc is carefully detached. The plate is then cut up so as to obtain the different names, and these are glued to handles.

XXIX.

WASTE AND ITS UTILIZATION.

There always will be certain quantities of waste in manufacturing articles from caoutchouc and gutta percha, no matter how carefully the work may be done, and, as the material is so very expensive, it is to the interest of the manufacturer to find ways and means of turning the waste to account.

The more we succeed in bringing the waste into such a shape that it can again be utilized in the same manner as purified caoutchouc, the more suitable will be the process by which this is achieved. A distinction must be made between the waste obtained from pure caoutchouc and that from the vulcanized mass.

Strict attention should therefore be paid in the factory to separate the waste, so that vulcanized pieces are not mixed with non-vulcanized.

There is but little difficulty in utilizing one kind of waste, but mixtures of odds and ends can only be applied to the lowest grades of manufacture.

The utilization of waste from crude caoutchouc is a very simple matter. It is only necessary to unite it to lumps and to pass them again through the rollers. The mass which is thus obtained frequently possesses a higher degree of plasticity than that originally used, as it has to be again rolled and the plasticity increases by a frequently repeated mechanical treatment.

For utilizing waste of vulcanized caoutchouc, it is comminuted as much as possible by a mechanical process—a grating apparatus or hollander is generally used for the purpose—and then mixed with pure caoutchouc, but the latter must be as finely divided as the vulcanite. While the materials are mixed, which is done by frequently rolling the heated mass, such a quantity of sulphur is added to them as may be sufficient to vulcanize the pure caoutchouc as much as the vulcanite.

The mass must be worked until it is so uniform that the separate parts cannot be distinguished from each other by the naked eye, and then articles may be fashioned from it in the usual manner, which are finally vulcanized by subjecting them to the burning process.

According to another process the waste is first cut into small pieces. These are boiled for several hours in caustic soda for the purpose of desulphurizing the caoutchouc. But this can only be done in a satisfactory manner when the pieces are very small, and the boiling is continued for a sufficiently long time.

In order to knead the waste after it has been boiled, together with caoutchouc, it is only necessary to continue the boiling until the pieces again possess the property of sticking together when heated.

It is claimed that by *Aco's* process, not only waste of ordinary vulcanized caoutchouc can again be utilized but also that of hard rubber. 100 kilogrammes (220 lbs.) of waste are treated in a closed vessel for two hours with a mixture of 10 kilogrammes (22 lbs.) of bisulphide of carbon, and 225 kilogrammes (495 lbs.) of spirit of wine. It is claimed that by this manipulation the mass becomes sufficiently softened to allow of its being mechanically treated by grating and kneading.

If this process could be used just as we have described, it would surpass all others in regard to simplicity. But the results of many experiments we have made in regard to this matter, have always been that many difficulties arise which can scarcely be overcome, and especially when hard rubber is mixed with the waste.

If the work is to be done by this process, it is at least necessary to sort the pieces of hard rubber from the waste and to treat them by themselves. But the simplest manner of utilizing the waste of hard rubber is the one we have already explained in a previous chapter, namely, to melt the waste and work it into varnish.

Another process of utilizing waste of vulcanized caoutchouc is as follows: The waste is comminuted as much as possible, and is then exposed to a temperature of 300° C. (572° F.) until a plastic mass has been formed. The heating is effected by steam, which is passed through a cylinder in which the small pieces of caoutchouc have been placed. Finally—

    Caoutchouc mass 5 kilogrammes (11 lbs.)
    Palm oil   60   grammes (2.1 ozs.)
    Sulphur   166   grammes (5.8 ozs.)

White lead or magnesia, chalk or oxide of zinc, 1 kilogramme (2.2 lbs.) are mixed together, and the articles formed from this mass subjected to the burning process.

Finally, we will mention *Newton's* method. This consists in treating the waste, which is placed in a well closed vessel, with camphene. Here it is allowed to remain for fourteen days, and is then heated to about 70° C. (158° F.) in order to complete the action of the camphene. When this has been done the greater part of the solvent is distilled off, and the tough mass remaining in the still is worked up. But if the vulcanite masses have been prepared with the assistance of metallic sulphides, the result is not satisfactory, and the action of the camphene continues for a long time when vulcanite is used which has been compounded with an extra large quantity of sulphur.

But no matter what process is used for preparing caoutchouc mass from waste of vulcanized or hard rubber, the product which is obtained never shows the properties of a prime article. It is, therefore, best to work the waste into hard rubber articles; of course, after it has been comminuted in a suitable manner and compounded with pure caoutchouc and sulphur, as the disparity of quality is not so much perceptible in these articles as in those from vulcanite masses, which must especially show a high degree of elasticity, no matter to what temperatures they may be exposed.

XXX.

ADULTERATIONS OF CAOUTCHOUC AND GUTTA PERCHA.

In a previous chapter we have drawn attention to the fact that the caoutchouc brought into commerce from the tropics frequently contains such large quantities of sand or earth that we must necessarily conclude that it has been done with dishonest intent. The object of admixing these bodies can easily be understood, namely, to increase the weight of the mass.

In describing the treatment of crude caoutchouc we have shown what a laborious and time-consuming labor it is to remove these impurities, and for this reason it may easily be comprehended why varieties which are known to be pure now bring a far higher price than those less pure. This may be the reason why for several years past a purer article has been brought to Europe, even from those regions which formerly were notorious for the impurities mixed with the caoutchouc, as the producers have very likely come to ·the conclusion, that it is more profitable for them to sell a smaller quantity by weight at a high price than a larger quantity of the impure article at a low price.

While impurities in caoutchouc can be easily detected by simply cutting it into pieces, this is a far more difficult matter in regard to gutta percha, and for this reason it is especially made the subject of many adulterations.

These consist not only of admixtures of foreign bodies—they could be easily detected—but of others, which are done in such a skillful manner, that frequently it is only possible to identify the spurious article by comparing it with undoubtedly pure gutta percha.

In fact, there is in the eastern part of Asia a special branch of industry devoted to the trade in the thickened juice of *getah malabeœya*. For a long time it was not known for what purpose this juice, which was brought into commerce by Chinese merchants from *Palembaug*, was used, until it was at last found that it was exclusively employed for adulterating gutta percha.

As gutta percha adulterated with getah was for

several years brought into the European market, and the admixture of this body was very injurious to the properties of the pure article, experiments were made to ascertain the properties of pure getah itself. As far as known they are as follows: It is very likely a mass obtained from the milky juice of a plant which is dried by artificial heat. Samples of getah, which came into the hands of European scientists, represented plates about 3 centimetres (1.17 inches) thick, of a grayish-brown color. They felt moist to the touch and possessed a certain degree of pliability, but this disappeared entirely as soon as the plates were thoroughly dried. Getah, like gutta percha, dissolves in chloroform, and therefore, if this is used as a solvent, it cannot be detected as an admixture of the pure article.

If treated with warm water it shows the same properties as gutta percha; it becomes soft and sticky. But there is an essential difference in the behavior of the two bodies towards boiling water. While pure gutta percha remains unaltered, even after long-continued boiling in water, getah is changed into an emulsion, that is, a milk-like fluid. By adding strong alcohol to this emulsion the getah is separated in a mass resembling bird-lime.

As has been mentioned, it is quite easily dissolved in chloroform, and only a small quantity of a black, insoluble body, very likely consisting of soot, remains behind. When treated with alcohol or ether getah parts with soluble bodies, namely, a wax-like body when the first is used, while that brought into solution, when ether is employed, is of a resinous nature.

The difference in the melting-points of the two substances is especially important as a means of recognizing and distinguishing getah from pure gutta percha. While the latter melts at 110°, or at the utmost 120° C. (230° or 248° F.), the melting-point of the former is as high as 170° C. (338° F.). Decomposition by dry distillation takes place at a correspondently higher temperature, dark-brown fluids appearing as products of decomposition, of the especial properties of which we thus far know nothing.

Experiments have also been made to work getah in the same manner as gutta percha, by cutting it up, washing, and forming it into a homogeneous mass by kneading between rollers. But the result of this labor was negative. The product obtained in this manner possessed a nearly black color, was not solid at an ordinary temperature, but of a consistency resembling that of glazier's putty, and had a very repulsive odor, which alone would prevent the use of this body for any purpose whatever.

There can be no doubt that the getah examined thus far also contains the constituents of the milky juice, which have become solid bodies, which can be removed by proper mechanical treatment in the same manner as the bad-smelling fluids, contained in the cavities of certain varieties of caoutchouc.

It is not easy to detect an admixture of getah with gutta percha. The latter, when mixed with it, does not show the lard-like character which distinguishes the pure article, but has a more spongy structure, exhibits a more grayish shade of color, and possesses a specific odor. While pure gutta percha smells as of leather and resin, it has a very disagreeable and peculiar odor when adulterated with getah, which helps an expert to detect the adulteration.

The determination of the melting-point of the mass is a simple and at the same time certain means of proving adulteration. Pure gutta percha becomes soft at a moderate temperature (48° to 50° C., 118.4° to 122° F.), and melts at 120° C. (248° F.), while the melting-point of the adulterated article is so much higher as immediately to strike one's attention.

The behavior of gutta percha towards chloroform may be used for determining the quantity of mechanical impurities mixed with any variety. A weighed quantity of the article to be examined is dissolved in chloroform. The vessel containing the mass is allowed to stand quietly for some time, in order to allow all solid bodies to settle on the bottom. The solution is then poured off, the residue rinsed off with some of the chloroform, dried, and weighed.

As caoutchouc is not quite as soluble as gutta percha, it becomes a matter of greater difficulty to determine the quantity of earth, sand, etc., mixed with it; and besides, such a determination would only be of secondary value, as in a large piece of caoutchouc many parts are entirely pure, while others are quite the reverse. An accurate knowledge of the commercial article will best protect the manufacturer against being imposed upon when buying the crude product.

XXXI.

EXAMINATION AND IMITATION OF CAOUTCHOUC AND GUTTA PERCHA COMPOSITIONS.

A COMPLETE chemical analysis is the only means of determining the constituents of a caoutchouc or gutta percha composition.

The analysis is commenced by burning a small quantity of the composition to be examined in a porcelain crucible, and heating the remaining mass until it is burned entirely white. This residue serves for the qualitative, as well as quantitative determination of the inorganic bodies contained in the composition.

Most of them always contain an admixture of a certain quantity of sulphur, and this may be determined in the simplest manner, by burning a weighed sample of the mass in a tube through which pure oxygen is conducted, and carrying the gases of combustion through water containing a small quantity of nitric acid.

The sulphur is changed into sulphurous acid and this is converted into sulphuric acid by the nitric acid, so that the fluid now contains sulphuric acid in solution. By adding some solution of baryta to the acid fluid, and drying and weighing the precipitate of sulphate of baryta, the quantity of sulphuric acid can be very nearly determined, and from this the quantity of sulphur, which originally was present, be calculated.

But the quantity of sulphur cannot be determined in this manner, if the mass contained metallic sulphides. In this case it must be divided into small pieces. These are gradually thrown into a crucible containing saltpetre which is heated to decomposition. A sharp detonation takes place every time a piece is thrown into the crucible, and finally the entire quantity of all the sulphur present is found in the melted mass in the form of sulphate of potassium. The contents of the crucible are dissolved in boiling water, a few drops of nitric acid added to it, and then some of a solution of chloride of barium, as long as a precipitate is formed. From the quantity of the dried precipitate of sulphate of baryta, the quantity of sulphur is calculated and the obtained figures serve for determining the quantities of sulphide of lead and trisulphide of antimony, which were contained in the original composition.

When the quantitative and qualitative composition of a caoutchouc composition has once been determined, it requires, as a general rule, only a certain number of experiments with compositions prepared according to the results of the analysis, to manufacture compositions possessing the same properties as the analyzed compound.

APPENDIX

STATISTICAL DATA RESPECTING THE CONSUMPTION OF CAOUTCHOUC AND GUTTA PERCHA AND THE PRODUCTION OF THESE INDUSTRIES.

It is always to the interest of the manufacturer to have a knowledge of the statistical data respecting products which are articles of general consumption, as from these the mercantile and industrial importance of the respective articles may be estimated. But the importance of gutta percha and caoutchouc for various industrial purposes has only been rightly appreciated within later years, and it may well be said that the objects for which these substances may be made useful are not all known at the present time.

Although caoutchouc has been known in Europe for nearly a century, it was originally but little used for industrial purposes and its production was also limited. But when the process of vulcanization became known, the rubber industry quickly became a considerable business, and the demand for the crude material increased in a corresponding manner. Since then, the reports of various countries, but especially those of England, have shown a remarkable increase in the importation of caoutchouc, which still continues at the present day.

The importation of gutta percha has increased in a similar manner, only in a far shorter time, as it is scarcely forty years since this body became known in Europe.

In the year 1862 the yearly production of caoutchouc was estimated as 8000 cwt., while 11 years later it amounted to 150,000 cwt., and at the present time may be estimated at 200,000 cwt.

Regarding the importation of crude caoutchouc and crude gutta percha we find in the following data the weight and commercial value of the imports.

England imported—

Crude Caoutchouc.

Year.	Cwt.	Value in pounds sterling.	
1869	136,421	1,134,585	($5,672,925)
1870	152,118	1,597,528	(7,986,640)
1871	161,586	1,630,262	(8,181,310)
1872	157,148	1,762,866	(8,814,330)
1873	154,491	1,719,383	(8,596,915)

Crude Gutta Percha.

1869	15,398	95,616	($478,080)
1870	34,514	196,951	(984,755)
1871	25,966	196,942	(984,710)

France imported—

Year.	Kilogrammes.	Cwt.	Average value per kilogamme in francs.	
1869	880,586	(17,611.61)	8.00	($1.60)
1870		(16,307.6)	8.00	(1.60)
1871		(22,572.34)	6.50	(1.30)

Hamburg imported—

Crude Caoutchouc.

Year.	Cwt.	Value in marks.	
1871	20,919	4,033,620	($1,008,405)
1872	29,162	6,394,395	(1,598,599)
1873	22,869	4,648,110	(1,162,028)

Crude Gutta Percha.

1871	1,293	200,685	($50,171)
1872	1,367	205,815	(51,454)
1873	1,961	304,820	(76,205)

The following figures indicate the value of manufactured articles.

Hamburg imported—

Rubber Shoes.

Year.	Cwt.		Marks.	
1871	3,355	valued at	606,450	($151,612)
1872	3,065	"	756,435	(189,109)
1873	3,938	"	906,920	(226,730)

Other Rubber Articles.

1871	10,438	valued at	3,622,725	($905,681)
1872	14,503	"	5,704,635	(1,426,159)
1873	16,670	"	6,590,310	(1,647,578)

Production of Rubber Goods in the United States.

States.	Establishments.		Hands employed.		Capital invested.		Value of products.	
	1860.	1870.	1860.	1870.	1860.	1870.	1860.	1870.
Connecticut......	9	13	809	1946	$1,265,000	$2,345,000	$2,276,000	$4,239,329
Maine............	..	1	....	9	......	4,000	......	31,500
Massachusetts....	5	16	298	1405	563,000	1,920,000	803,000	3,183,218
Missouri.........	..	1	....	4	......	2,000	......	4,500
New Jersey.......	5	12	817	807	870,000	1,034,000	1,333,000	2,248,300
New York........	7	10	757	1008	775,000	1,777,000	1,127,750	3,076,720
Pennsylvania.....	1	1	8	1	5,000	800	12,000	1,400
Rhode Island.....	2	2	113	845	156,000	403,000	246,700	1,804,868
	29	56	2802	6025	3,634,000	7,485,800	5,798,450	14,589,835

Among the products of 1870 were :—

    1,250,000 lbs. of car springs
      906,000 lbs. of belting and hose
      552,500 dozen of braces
    5,402,666 pairs of boots
       30,000 coats.

There were consumed :—

    8,413,320 lbs. of caoutchouc
    2,934,575 yards of cloth
    2,391,451 lbs. of cotton
       2,900 lbs. of silk.

The value of all the materials was $7,434,742.

The imports into the United States of caoutchouc and gutta percha for 1881, were 20,015,176 lbs. of the value of $11,054,949; for 1882, 22,712,862 lbs. of the value of $14,264,903.

The Manufacture of Rubber and Elastic Goods, and Rubber Vulcanized in the United States, according to the Census of 1880, was as follows:—

Manufacture of	Number of Establishments.	Capital.	Hands Employed.			Wages paid	Materials.	Products.
			Males.	Females.	Children and Youths.			
Rubber and Elastic Goods	89	$5,987,987	3658	2281	294	$2,283,874	$9,138,537	$13,609,552
Rubber vulcanized . . .	3	226,200	335	150	10	154,700	391,200	767,200
Total	92	$6,214,187	3993	2431	304	$2,438,574	$9,529,737	$14,376,752

SUBSTITUTES FOR CAOUTCHOUC AND GUTTA PERCHA.

Notwithstanding the increase in importation, the price of the raw materials remains at a comparatively high figure, and many experiments have been made to find a body which possesses the same properties as catouchouc or gutta percha, but that could be produced at a cheaper rate.

Although many of these compositions are as well adapted for certain purposes as caoutchouc and gutta percha, yet, generally speaking, the experiments have not furnished satisfactory results, and we know of no substance which, respecting its chemical indifference, even approaches caoutchouc or gutta percha.

Neither has there been success in imparting to the masses recommended as substitutes for caoutchouc, the extraordinary elasticity which makes this body indispensable for the manufacture of many articles. What has been made public about this has never been introduced to any extent into the industry; the best proof that the respective statements were only of secondary value. For this reason it is only necessary to discuss briefly the substitutes for caoutchouc and gutta percha, and the more so, as the compositions of any value, but which must always contain a certain quantity of caoutchouc or gutta percha, have already been treated in a previous chapter.

Manufacturers had placed great hopes in the products which could be prepared from linseed oil, as it is well known that masses can be produced from it, which, to a certain extent, equal caoutchouc compositions in regard to elasticity and toughness.

It is well known that the greater part of the fluid designated as siccative, or oil varnish, is prepared from linseed oil, by boiling it with compounds of lead, such as litharge, minium, or sugar of lead. Linseed-oil varnish, when correctly prepared, should be clear, and dry in a few hours to a transparent, glassy mass of great tenacity. By changing the mode of preparing linseed-oil varnish in so far as to boil the oil until it is thickly fluid, and spins threads when it is taken from the boiler, a mass is obtained which in drying assumes a character resembling that of a thick congealed solution of glue.

Resins are added to the hot mass—the quantity depends on the character of the product to be obtained—to give it a consistency as nearly as possible resembling caoutchouc. There are many receipts for such masses, but they all agree in the principal points, namely, the mixing of the linseed oil with resins, the difference being in the quality as well as quantity of the resins or resinous bodies to be used.

Shellac, colophony, or ordinary pitch may also be used, but it is advisable to always add some rosin oil, in order to diminish the brittleness the mass generally acquires from the addition of the resins.

Among the best of these compositions is one consisting of boiled linseed oil and shellac. This is prepared by boiling the oil until the hot mass shows the consistency of thick glue. Unbleached shellac and some lamp-black are then stirred into it, and the boiling is continued until the shellac has become intimately mixed with the oil. The mass is put into flat vessels, where it is allowed to congeal. It is then taken from the vessels, and the blocks, while still warm, are repeatedly passed between rollers for the purpose of mixing the constituents as intimately and homogeneously as possible.

It has frequently been stated that this mass could be

vulcanized, and that the product obtained by this process would show the same properties as vulcanized caoutchouc. We have repeatedly tried to bring about such a vulcanization, but have never been able to obtain any satisfactory results. Only in such cases where a certain quantity (between 15 and 25 per cent.) of caoutchouc had been mixed with the mass was it possible to vulcanize it, but even in this case we have good reason to suppose that only the caoutchouc and gutta percha became actually vulcanized, while the mass consisting of linseed oil and shellac remained unaltered.

By mixing powdered wood, cocoa-nut shells or fibres with this mass, substances are obtained which are quite suitable for making ornamentations, boxes, match-cases, etc. The so-called linoleum, which is very durable, is said to be manufactured from a mass consisting of boiled linseed oil, resin, and finely ground cork. The mixture, while hot, is spread upon a tissue which serves as a base for the whole.

For certain purposes the masses prepared from linseed oil may also be used as a basis for a coat of caoutchouc or gutta percha. Plates are prepared from the mass, and these, which are covered on both sides with a thin plate of caoutchouc or gutta percha, are then passed through hot rollers and united in one sheet. In this manner articles can be prepared a great deal cheaper than from the pure materials.

The surfaces of hollow articles, which have been formed from this mass, by pressing them in suitable forms, may be coated with caoutchouc by applying a solution of the latter to them, and may then be vulcanized by the burning process.

This process is well adapted for manufacturing large articles which would be too expensive if prepared from the pure materials; such, for instance, as bath-tubs, pitchers, vases, small pieces of furniture, figures, and a number of other articles.

SUBSTITUTE FOR GUTTA PERCHA.

This substance is gained by boiling the bark of the birch tree, especially the outer parts, in water over an open fire. A black fluid mass, but which quickly becomes solid and very compact on exposure to the air, remains behind in the evaporating vessel.

This mass possesses all the properties of gutta percha and may be used for the same purposes. It has even the advantage over the latter of not becoming full of cracks when exposed to the air, and of being more solid and costing less to prepare. The inventor has named this new product "*French Gutta Percha*," and of course does not claim the sole right of preparing it, as any kind of apparatus may be used, but has patented its application to industrial purposes.

A particular kind of caoutchouc which is very solid and comparatively inexpensive, is obtained by mixing 55 parts of it with 45 parts of French gutta percha.

L. Dankwerth and *F. Sanders* of *St. Petersburgh* have recently patented a mass which they claim to be a perfect substitute for caoutchouc and gutta percha and which may be used for insulating telegraph wires, etc. They also claim it to be fully as elastic and tough as caoutchouc, and less sensitive to exterior influences, and that it cannot be injured by great pressure and a high temperature. It is prepared in the following manner: 1 part by weight of a mixture of equal parts of wood tar, oil, and coal tar oil, or, of the latter alone, is heated for several hours at 140° to 150° C. (252° to 270° F.) with 2 parts by weight of hemp-oil until the mass can be drawn into threads. Then $\frac{1}{3}$ part by weight of linseed-oil somewhat thickened by previous boiling, is added, and to every 100 parts of this composition from $\frac{1}{20}$ to $\frac{1}{10}$ part of ozokerite with some spermaceti. It is then again heated to the same temperature, and then, finally, $\frac{1}{15}$ to $\frac{1}{12}$ part of sulphur is added to it. The product is worked up in the same manner as caoutchouc.

Cement for Vulcanized Caoutchouc.

	Parts.
Stockholm pitch,	3
American rosin,	3
Oil of turpentine,	8
Caoutchouc,	6
Petroleum,	12

are heated and stirred together. Should the mixture be too thick for certain purposes, it may be thinned by adding more oil of turpentine.

The surfaces to be joined together should first be roughened by rubbing them with emery or pumice stone before the cement is applied.

NEW RUBBER COMPOUND.

A hard and metal-like product (possessing some of the characteristics of lead and some of rubber, and yet differing from either) especially suitable for bearings, or bearing-linings, stereotype plates, etc., may be made by combining —

	Parts.
Caoutchouc,	2
Sulphur,	1
Plumbago,	4 to 5

duly mixed together, in a heater in the manner usual with rubber compounds, next moulding the articles desired, and then vulcanizing at a temperature of 155° to 160° C. (311° to 320° F.) for seven or eight hours. By adding from a half to one part of whiting to the above, a compound suitable for tool handles and similar articles may be formed, and, with a larger quantity of plumbago, sticks or rods suitable for carpenters' pencils may be made.

INDEX.

Aco's process for utilization of waste, 250
Adulterations of caoutchouc and gutta percha, 252–257
African caoutchouc, 39, 40
Air, effect of, on caoutchouc, 45
Albane, 156, 157
Alcohol, absolute, behavior of caoutchouc toward, 48
    treatment of caoutchouc by, 44
American caoutchouc, 37, 38
    more easily worked than others, 57
Apocyneæ, 32
Apparatus for coating wires with gutta percha, 221
    for kneading caoutchouc solutions, 135, 136
    for protecting workmen from the effects of vapors, 236–238
    for spreading caoutchouc mass upon tissues, 234
    for treating caoutchouc with chlorine, 116–117
Artificial ivory, 114–120
Artocarpaceæ, 32
Asphaltum cement, 146

Bahea gumnifera, 32
Balata and Coorongit, 30
Balenite, 124, 125
Bark of the birch tree a substitute for gutta percha, 268
Behavior of caoutchouc in heat, 54–56
Behavior of caoutchouc—
    toward solvents, 47–54
    toward sulphur, 46
Belts, machine, composition for, 172–174
Benzol, a solvent of caoutchouc, 49
    for preparing caoutchouc solution, 134
Birch bark a substitute for gutta percha, 268
Bisulphide of carbon, action of, upon vulcanized caoutchouc, 72
    and alcohol solvent, how prepared, 52
    a solvent of caoutchouc, 49
    for preparing caoutchouc solutions, 134
Bleaching caoutchouc, 115–118
    effect of, 115, 116
    of gutta percha, 167–171
Bornesite, 44
Brockedon's experiments on the heat developed in the sudden extension of caoutchouc, 42
Burning apparatuses, 83, 84
Burning hard caoutchouc, 106, 107
    in vulcanization, 74, 75
    of hard rubber, temperature for, 108
    of the caoutchouc and the sulphur mass, 78–86
    operation, 84–86

Cables, Miller's examination of, 159
   submarine, length, etc., of some now in use, 223
Calenders, 189–191
Caoutchouc, African and Madagascar, 39, 40
   American, 37, 38
   a hydrocarbon, 44
   and gutta percha, behavior of, almost identical, 30
      composition of, for machine belts, 172–174
      compositions, examination and imitation of, 257, 258
      consumption of, 259–264
      plates, 184–193
      working, 180–183
   and sulphur mass, burning the, 78–86
      mechanical combination of, 76–78
   bleaching, effect of, 115, 116
   by whom first made known, 26
   Carthagena, 37
   cements, 141–144
   coloring, 97–101
   commercial, 37–40
   compositions, 120–132
      water-proof tissues by means of, 242–244
   consists of an insoluble base with pores containing soluble parts, 50
   containing earth and sand, treatment of, 36
   crude, working, 56–67
   different modes of preparing the raw material, 36
   East Indian, 39

Caoutchouc—
   enamel, 130–132
   first knowledge of, derived from South America, 26
   how classified, 37
   how obtained from the trees, 34
   how treated in Central America, 35
   industrial utilization of, 27
   lacquers, 132–149
   leather, 123, 124
   oil of, 52
      bodies present in, 54, 55
   Para, 38
   parts dissolved in certain solvents, 49
   plants, cultivation of, 28
      which yield it in large quantities, 31–33
   present mode of gathering, 35
   properties of, 40–56
   properties of certain solutions of, when evaporated, 49, 50
   solutions, 133
   sponges, fabrication of, 223–225
   value of, on what it depends, 37
   vulcanized, cement for, 269
   West Indian, 38
   what it is, 31
   whence derived, 26
   where found and how obtained, 31–36
   where obtained, 28
Carthagena caoutchouc, 37
Castilion elastica, 32
Cecropia peltata, 32
Cement, elastic gutta percha, 145
   for glass, 141, 142, 144
   for leather, 144
   for rubber combs, 145

INDEX. 273

Cement—
    for rubber shoes, 142
    for vulcanized caoutchouc, 269
    hard, 144
    soft, 143
Cements, caoutchouc, 141–144
    gutta percha, 144–147
Chemical properties of caoutchouc, 44–46
    of gutta percha, 155–160
Chloride of sulphur, preparation of, 91–93
    vulcanization with, 87–93
Chlorine, treating caoutchouc with, 116
Chloroform, dissolving caoutchouc in, 117
Cleansing of crude gutta percha, 161–165
Coating of wires with gutta percha, 219–223
Cohesion of the particles of caoutchouc, 43
Colophony as a substitute for caoutchouc, 266
Coloring caoutchouc, 97–101
    hard rubber, 108
Coloring substances for hard rubber, 110
Comandine one of the earliest discoverers of caoutchouc, 27
Commercial caoutchouc, 37–40
Compositions, examination and imitation of, 257, 258
    gutta percha, 171–180
    hard gutta percha, 174–176
    of gutta percha and wood, 176–177
Compound, new rubber, 269, 270
Consumption of caoutchouc and gutta percha, 259–264
Coorongit and Balata, 30
Cords, square, from crude caoutchouc, 196–199

Cords, square—
    from vulcanized caoutchouc, 199, 200
Cornite, 104
Crude caoutchouc, working, 56–67
Cutting caoutchouc, machine for, 58, 59
    plates from blocks and cylinders, 184–187

Dambonite, 44
Dankwerth and Sanders' patent substitute for gutta percha, 268
De Colbry's vulcanizing process, 87
Deodorization of waterproof fabrics, 240–242
Deodorizing vulcanite, 101, 102
Desulphurized rubber, 73
    how produced, 73
    vulcanite, 102–104
Dissolution of caoutchouc, complete, requires manipulation, 50
Dumas' thin plates for waterproofing, 228
Dusting hard rubber, 109

East Indian caoutchouc, 39
Ebonite, 114–120
    American, receipt for, 120
Eburite, 114–120
Elasticity of caoutchouc, and on what it depends, 41–43
    what it depends upon, 50
    temperature effect on, 41, 42
Elastic webbings, fabrication of, 245–247
Enamel, caoutchouc, 130–132
Enamelling hard rubber, 109
England, imports of caoutchouc and gutta percha into, 260
Erasing rubber, 148

24

274 INDEX.

Ether, action upon vulcanized caoutchouc, 72
　a solvent of caoutchouc, 49
Euphorbiaceæ, 32
Examination and imitation of caoutchouc and gutta percha compositions, 257, 258
Expansion of caoutchouc in certain fluids, 47

Ficus elastica, 32, 33
Fluavile, 156, 157
France, imports of caoutchouc and gutta percha into, 260
French gutta percha, 268
Fresneau, one of the earliest discoverers of caoutchouc, 27
Fry's patent method for obtaining solutions for coutchouc and gutta percha, 53

Gerard's observations on the elasticity of caoutchouc, 41
Getah malabeœga, adulteration of gutta percha with, 253
　　detecting, 254–256
　　properties of, 254
Glue, marine, 138–140
Goodyear, discovery of vulcanizing caoutchouc by, 69
　zinc penta-sulphide recommended by, 105
Grinding and polishing compositions, 127–130
Guatemala caoutchouc, 37
Guayaquil caoutchouc, 37
Gutta, abane and fluavile, relations between, 157
　percha, 150
　　and caoutchouc, behavior of, almost identical, 30
　　composition for machine belts, 192–194

Gutta percha and caoutchouc—
　consumption of, and production of the industries, 259–264
　plates, 184–193
　working, 180–183
　behavior of, at different temperatures, 153, 154
　bleaching of, 167–171
　cements, 144–147
　chemical properties of, 155–160
　Clark's experiments on, 158, 159
　cleansing of crude, 161–165
　composition, 171–180
　compositions of, 176–177
　Dankwerth and Sanders' patent, 268
　French, 268
　identity of, 28
　industry, early introduction of, 29
　physical properties of, 152–155
　properties of, 152–160
　solvents for, 155
　the plant from which it is derived, 150
　vulcanization of, 165–167
　whence derived, 26
　when first made known, 25
　where found, 150
　pure, 156

Hamburg, import of rubber goods into, 261
　imports of caoutchouc and gutta percha into, 261
Hancornia speciosa, 32

INDEX. 275

Hard gutta percha compositions, 174–176
  rubber, 69
    coloring substances for, 110
    contraction of, in burning of, 107
    elastic and pliable proportion of constituents of, 111–113
    hardness and elasticity of, on what they depend, 111
    indifference to chemical influences, 113
    lacquer, 140, 141
    materials employed for making, 105
    preparation of, 104–113
    uses of, 104
    with great hardness, proportion of constituents for, 112, 113
Heat, behavior of caoutchouc in, 54–56
  developed in the sudden extension of caoutchouc, 42
  necessary in vulcanization, 78–80
  the dissolving powers of fluids increased by, 134
Heveenoid, 147, 148
Hollander, the, 61–63
Hollow bodies, moulding, 214–223
Horses' hoofs, gutta percha cement for, 146
Hose, caoutchouc, 205, 206
  gutta percha, 209–214
  ordinary, 206, 207
  or tubing, caoutchouc and gutta percha, 205–214
  with inclosures, 208, 209
Hydriodic acid, treatment of caoutchouc with, 44

Imitation of caoutchouc and gutta percha compositions, 257, 258
Inosandra gutta, 150
Ivory, artificial, 114–120

Jeffery's marine glue, 140
Jintawan, 30

Kamptulicon, 121–123
Keratite, 104
Kneading caoutchouc solutions, 135, 136
  rollers, 63–65

Lacquers, caoutchouc, 132–149
Leather straps, cement for, 146
Light, effect of, on caoutchouc, 45
Linseed oil as a substitute for caoutchouc, 265, 266
  boiled, and shellac, substitute for caoutchouc, 266
  dissolving caoutchouc in, 51, 52
Lüdersdorf's treatment of caoutchouc with sulphur, 69

Machine belts, composition for, 172–174
  for cutting crude caoutchouc, 58–60
Machines for the treatment of caoutchouc, 66
Mackintosh, 227, 228
Mackintosh's improvement, what it consists of, 229
Madagascar caoutchouc, 39, 40
Marine glue, 138–140
  for damp walls, 138, 139
Massive articles, forming or moulding, 214–223
Matezite, 44
Mazer wood and gutta percha, identity of, 28

Mechanical combination of caoutchouc with sulphur, 76–78
Metallized caoutchouc, 148
Microscopic examination of caoutchouc, 41
  of gutta percha, 154
Milk-weed, caoutchouc in, 31
Milky juices of plants, what they contain, 26
Montgomery, Dr. Wm., first to import gutta percha into England, 28, 29
  investigations in regard to gutta percha in the East Indies, 29
Moulding of massive articles and hollow bodies, 214–223

New Granada, caoutchouc from, 37
Newton's process of utilizing waste, 251

Oxygen, effect of, on caoutchouc, 45

Para caoutchouc, 38
Parke's cold vulcanizing process, 87
Payen's analysis of gutta percha, 156
Petroleum, refined, for treating caoutchouc, 70
  refining of, 93
Physical properties of caoutchouc, 40–43
Pitch as a substitute for caoutchouc, 266
Plants, principal families of, which furnish caoutchouc, 32
Plastite, 125–127
Plates, cutting, from blocks and cylinders, 184–187
  manufacture of, from solutions, 187–189
  preparation of, by rolling, 189

Plates, preparation of—
  from vulcanite masses, 193–195
Plating, hard rubber, 109
Polishing and grinding compositions, 127–130
Priestley on the application of caoutchouc, 27
Properties, chemical, of caoutchouc, 44–46
  of caoutchouc, 40–56
  physical, of caoutchouc, 40–46

Raw caoutchouc, want of uniformity, 56
Resin, addition of, to caoutchouc, 100
  in caoutchouc, 44
Rolling, preparation of plates by, 189
Rousseau's solutions of gutta percha, 179, 180
Rubber compound, new, 269, 270
  shoes, fabrication of, 225–227

San Salvador caoutchouc, 37
Shavings of caoutchouc, passing through the hollander, 62, 63
  to obtain thin, 60
  washing, 61
Shellac as a substitute for caoutchouc, 265
Shoes, rubber, fabrication, of, 225–227
Siphonia elastica, 32, 33
Solutions, caoutchouc, 133
  for obtaining the best, 49
  gutta percha and caoutchouc, by Fry's patent method, 53
  manufacture of plates from, 187–189

Solvents, behavior of caoutchouc toward, 47–54
   for caoutchouc, 48
      the best, for practical purposes, 52
Sorel's gutta percha compositions, 177–179
Specific gravities of different samples of caoutchouc, 43
Sponges, caoutchouc, fabrication of, 223–225
Spreading caoutchouc mass upon tissues, 233–239
Stamps, caoutchouc, 247, 248
Statistical data of the consumption of caoutchouc and gutta percha, and the production of the industries, 259–264
Stretching machines, 191–193
Submarine cables, length, etc., of some now in use, 223
Substitutes for caoutchouc and gutta percha, 265–270
Sulpho-metals for vulcanizing, 94–104
Sulphur, absorption of, by caoutchouc, 73, 74
   and caoutchouc mass, burning the, 78–86
   and caoutchouc, mechanical combination of, 76–78
   behavior of, toward caoutchouc, 46
   chloride of, and bisulphide of carbon for treating caoutchouc, 76
   vulcanization with, 87–93
   effects of excess of, in vulcanite, 72, 73
      on the elasticity of articles, 72, 73
   mechanical combination of, for treating caoutchouc, 70
   proportion of, in forming vulcanite, 72

Sulphur—
   quantity used in making hard rubber, 105
   various modes of introducing into caoutchouc, 70
   vulcanizing with pure, 73–86
Sun, effect of, on caoutchouc, 45

Temperature, effect of, on the elasticity of caoutchouc, 41, 42
Threads, gutta percha, 203–205
   preparing caoutchouc and gutta percha, 196–205
   round caoutchouc, 200–203
Tissues, waterproof, fabrication of, with caoutchouc, 227–242
   with caoutchouc compositions, 242–244
   with caoutchouc inclosures, 239, 240
Turpentine, oil of, a solvent of caoutchouc, 49
   recommended as a solvent, 51

United States, imports of caoutchouc and gutta percha into, 264
   production of rubber goods in, 262, 264
Urceola elastica, 32, 33
Utilization of waste, 248–252

Varnish for gilders, 138
   for glass, 138
   for leather, 137
Vulcanite, 67–73
   coloring, 97–101
   deodorizing, 101, 102
   desulphurized, 102–104
   masses, 97–101
      preparation of plates from, 193–195

Vulcanite—
    proportion of sulphur in making, 72
        necessary in forming, 72
    pure, 97
    treating with solvents, 72
    treatment of with ether, 71
Vulcanization, degrees of heat required in, 78-80
    of gutta percha, 165-167
    with chloride of sulphur, 87-93
Vulcanized caoutchouc, cements for, 269
    elasticity of, 71
    preparation of, 67-73
    properties of, 71-73
    substances mixed with caoutchouc in preparing, 70
    water absorbed by, 71
Vulcanizing by plunging into melted sulphur, 73, 74
    Lüdendorf's and Goodyear's discoveries, 69
    properties given to caoutchouc by, 68, 69
    with pure sulphur, 73-86

Vulcanizing—
    with sulpho-metals, 94-104
Waste and its utilization, 248-252
    of hard rubber, utilization of, 111
    of vulcanized caoutchouc, 249
Water, absorption of, by caoutchouc, 48
    caoutchouc not dissolved in, 47
Waterproof tissues, by means of caoutchouc compositions, 242-244
    fabrication of, with caoutchouc, 227-242
Webbing, elastic, fabrication of, 245-247
West Indian caoutchouc, 38
Wires, coating of, with gutta percha, 219-223
Wood and gutta percha, compositions of, 176, 177
Working caoutchouc and gutta percha, 180-183

Zinc penta-sulphide, 105

www.ingramcontent.com/pod-product-compliance
Lightning Source LLC
Chambersburg PA
CBHW032006230426
43672CB00010B/2267